Game Theory

Outsmart This Shit

Igor Laplace

Igor's Fucking Awesome Math Books

Game Theory: Outsmart This Shit
ISBN: 9781779660749

Copyright © 2024 Igor Laplace.
All rights reserved. No part of this publication may be reproduced, distributed, or transmitted in any form or by any means, including photocopying, recording, or other electronic or mechanical methods, without the prior written permission of the publisher, except in the case of brief quotations embodied in critical reviews and certain other noncommercial uses permitted by copyright law. For permission requests, write to the publisher, addressed "Attention: Permissions Coordinator," at the address below.

IGOR'S FUCKING AWESOME MATH BOOKS

Published by:
Igor's Fucking Awesome Math Books
Paris, France

All publishing rights owned by Madison Matti Charlton.

Contents

Introduction   1
Understanding Game Theory   1

Decision-Making in Game Theory   33
Rationality and Decision-Making   33

Types of Games   67
Two-Player Zero-Sum Games   67
Solving Sequential Games: Backwards Induction   94
Cooperative Games   101

Game Theory in Social Networks   135
Social Network Analysis and Game Theory   135

Game Theory and Evolutionary Biology   175
Evolutionary Game Theory   175

Strategic Interaction and Game Theory   209
Bargaining and Competitive Strategy   209

Game Theory in Financial Markets   241
Game Theory and Financial Decision-Making   241

Game Theory in Artificial Intelligence   277
Machine Learning and Game Theory   277

Index   309

Introduction

Understanding Game Theory

What the fuck is Game Theory?

Game Theory is a concept that might sound intimidating, but don't worry, I'm here to break it down for you. So, what the fuck is Game Theory all about? Well, Game Theory is a branch of mathematics that helps us understand strategic decision-making in situations where an individual's outcome depends on the choices of others.

Think about it like this: life is a game, and we're all players making strategic moves to get what we want. Whether it's deciding what to have for dinner or negotiating a business deal, we're constantly faced with choices that are influenced by the actions of those around us. Game Theory provides us with a set of tools to analyze and predict these interactions.

But Game Theory is not all serious business. It's like a blend of chess, poker, and The Hunger Games. It's about outsmarting your opponents, predicting their moves, and finding the best strategies to achieve your goals. It's a bit like being a motherfucking genius detective, trying to unravel the mysteries of human behavior.

So, why the fuck should you care about Game Theory? Well, understanding Game Theory can help you navigate through complex decision-making situations. It can help you become a strategic badass in both your personal and professional life. By breaking down situations into games, you can analyze the possible outcomes, assess risks and rewards, and choose the best course of action to maximize your gains.

Game Theory isn't just some nerdy-ass theory confined to academia. It applies to various fields like economics, politics, biology, psychology, and even fucking online gaming. It's like a versatile Swiss army knife that has applications everywhere. Whether you want to understand how companies compete, solve conflicts, or analyze social networks, Game Theory is here to help.

Now, don't be fooled into thinking that Game Theory is all about winning at any cost. It's not just about being a ruthless asshole. Ethics play a crucial role in Game Theory too. We explore questions like fairness, cooperation, and the moral implications of different strategies. So, you can be both a rational and ethical player in the game of life.

But let's be real here. Game Theory isn't just about playing fair and being all goody two shoes. Sometimes, you gotta bend the rules a little. Cheating and deception can be part of the game. Just like in relationships, politics, and business, people sometimes play dirty. Game Theory helps us understand these behaviors and assess the consequences. So, Game Theory ain't just for saints - it's for badasses too.

With the rise of digital technology, Game Theory has also evolved. Online gaming and virtual economies have introduced new dimensions to strategic decision-making. Thinking about how you can gain an advantage, whether through leveling up, getting more in-game currency, or trading with other players, requires a Game Theory mindset. So, even gaming is an opportunity to learn some real-life skills.

Now, here's a mindfuck for you. Imagine the future of Game Theory. Picture a world where Artificial Intelligence (AI) and Machine Learning are the players. Yep, AI is learning Game Theory to beat us at our own game. Understanding how AI and human players interact is becoming increasingly important. We'll explore these concepts in future chapters, so stay tuned.

To sum up, Game Theory is about making sense of the strategic choices we face in life. It's about understanding the dynamics of decision-making, predicting the behavior of others, and finding the best strategies to achieve our goals. Whether you want to be a strategic badass in business, politics, social networks, or even fucking AI, Game Theory has got your back.

So, let's dive deep into the fascinating world of Game Theory and learn how to play smart, not hard. Get ready to level up your decision-making skills and kick ass at the game of life!

Evolution of Game Theory

Game Theory didn't just pop out of nowhere, folks. It has a rich history and has been evolving over time, just like your taste in music or your choice of fashion. In this section, we're going to take a wild ride through the evolution of Game Theory and see how it has shaped up to become the bad-ass field it is today.

Back in the day, in the early 20th century, some bright minds like Emile Borel and Ernst Zermelo started laying the foundation for Game Theory. They were interested in figuring out how multiple players interact in a strategic setting and what kind of strategies they should adopt to maximize their gains.

But it wasn't until the 1940s that a true pioneer named John von Neumann stepped onto the scene and made Game Theory his bitch. He published a groundbreaking book called "Theory of Games and Economic Behavior" (yeah, they really knew how to make catchy titles back then) where he formalized the field and introduced the concept of a "game" in a mathematical sense.

You see, Game Theory is all about representing strategic interactions between rational decision-makers as games. No, no, not board games or video games, but games that involve players making choices and facing consequences based on the choices of others. Von Neumann's book laid out the basic principles of Game Theory and set the stage for further exploration.

But the party didn't stop there. In the 1950s, two dudes named John Nash and Reinhard Selten crashed the Game Theory scene with their contributions to the field. Nash introduced the concept of "Nash equilibrium," which is a fancy way of saying that players have reached a state where nobody has an incentive to change their strategy. It's like a Mexican standoff, but without the sombreros and ponchos.

Selten, on the other hand, delved into the notion of "subgame perfect equilibrium," which is a way of analyzing sequential games where players make choices at different stages. He was all about finding strategies that are so badass that no player can screw things up, no matter how hard they try.

As Game Theory kept evolving, more and more players joined the party. The 1970s saw the rise of Robert Axelrod, who organized the mother of all tournaments called the Prisoner's Dilemma tournament. This guy invited a bunch of strategies to duke it out in the infamous

Prisoner's Dilemma game, and the winner turned out to be... drumroll, please... "Tit-for-Tat"!

"Tit-for-Tat" is one cunning strategy that starts off nice and cooperative but then responds to its opponent's previous move. If you're nice to it, it'll be nice to you. If you screw it over, it'll screw you right back. It's like the ultimate revenge plot, but in the realm of Game Theory.

Fast forward to the 21st century, and Game Theory has become hotter than ever. With the rise of digital technologies and online gaming, Game Theory has found its way into virtual worlds and economies. Players can now strategize and make decisions in video games, online auctions, and cryptocurrency markets. It's like a game within a game within another game, and it's mind-blowing!

But wait, there's more! The future of Game Theory holds even more excitement. With the advent of Artificial Intelligence (AI) and Machine Learning, we can expect Game Theory to merge with these fields and create some mind-boggling applications. Imagine advanced AI agents negotiating, collaborating, and competing with each other using Game Theory. It's a whole new level of awesomeness.

So there you have it, the evolution of Game Theory in a nutshell. From the early pioneers like von Neumann to the modern-day fusion with AI, Game Theory has come a long way. It's a field that continues to captivate and challenge us, revealing the intricate ways in which we strategize and interact. So buckle up, my fellow game theorists, because the game is just getting started.

Importance of Game Theory in Various Fields

Game theory is like the holy fucking grail of decision-making. It isn't just for nerds and mathematicians; it has pervasive applications across a wide range of fields. Let's dive into some of the key areas where game theory holds its weight.

Economics

Game theory is a game-changer in economics. By modeling strategic interactions between individuals and firms, it enables economists to understand and predict market dynamics, pricing strategies, and other important economic phenomena.

One classic example is the Prisoner's Dilemma, which highlights the tension between cooperation and self-interest. This dilemma helps economists understand why two rational individuals might not cooperate even when it's in their mutual interest. Game theorists have used this concept to analyze cartel behavior in industries like oil and drugs (not the happy kind, sorry).

Another essential concept is Nash equilibrium, named after John Nash, the guy from "A Beautiful Mind." It's all about finding stability in strategic interactions. This equilibrium is achieved when no player has an incentive to unilaterally deviate from their chosen strategy. In economics, Nash equilibrium is used to analyze oligopolistic markets, where a few powerful firms control the market supply.

Political Science

Game theory is like the Rosetta Stone for understanding political behavior. It helps us analyze how politicians make decisions, engage in negotiations, and strategize in competitive environments.

Take voting theory, for example. Game theorists have developed mathematical models to uncover the strategies behind voting mechanisms. Ever wonder why some politicians are more likely to pander to particular interest groups? Well, game theory can shed some light on that too.

Game theory is also used to study international relations. It helps explain how countries interact and make decisions in contexts like arms races, trade negotiations, and climate agreements. It provides insights into the strategies countries employ to maximize their self-interest without compromising global stability.

Biology

Game theory isn't just about numbers and money; it's also critical in understanding the living world around us. Evolutionary biology, in particular, benefits greatly from game theory.

Evolutionary game theory allows scientists to understand the dynamics of natural selection and the evolution of behaviors. It helps explain why certain traits, like cooperation or aggression, persist in a population over time.

One famous concept in this field is the Hawk-Dove game, which models the strategic interactions between animals competing for resources. It illustrates the balance between risk-taking and aggression in conflicts. By extending this concept to social interactions and cooperation, game theory has enabled us to study the evolution of complex behaviors in animals, like reciprocal altruism.

Computer Science

In the digital age, game theory has made its way into computer science like a boss. It's used in designing intelligent algorithms, multi-agent systems, and even cybersecurity.

Reinforcement learning, a popular technique in machine learning, is heavily influenced by game theory. It helps create algorithms that make decisions based on past experiences and feedback from the environment.

In multi-agent systems, game theory is used to study the interactions between different intelligent agents. It helps us understand how agents can coordinate, negotiate, and make decisions in complex environments.

Game theory is also crucial in cybersecurity. By modeling attacks and defenses as a strategic game, we can develop robust security measures to combat cyber threats (take that, hackers!).

Social Sciences and Psychology

Game theory has found its place in various social science disciplines, including sociology, psychology, and anthropology.

In sociology, game theory allows us to study social networks, cooperation dynamics, and collective action problems. It helps us understand why individuals cooperate in certain situations and how social norms, trust, and reputation influence our decisions.

Psychologists have embraced game theory to study human behavior in experimental settings. It provides a framework to model decision-making, strategic thinking, and even moral reasoning. Game theory experiments have provided insights into human biases, heuristics, and the role of emotions in decision-making.

Now, game theory isn't a crystal ball that magically predicts our every move. It's a tool, a way of thinking, that helps us better understand strategic interactions and make informed decisions. So, whether you're an economist, a politician, a biologist, or just a curious motherfucker, game theory will challenge your assumptions, broaden your perspective, and help you play smart, not hard.

Resources

To dive deeper into game theory and its applications, here are some kick-ass resources for you to check out:

- **Books**:
    - "Game Theory: An Introduction" by Steven Tadelis (love this book, highly recommend)
    - "The Strategy of Conflict" by Thomas C. Schelling (oldie but goldie)
    - "Evolution and the Theory of Games" by John Maynard Smith (must-read for biology enthusiasts)
    - "Thinking Strategically" by Avinash Dixit and Barry Nalebuff (great insights on strategic decision-making)

- **Online Courses**:
    - Coursera: "Game Theory" by Stanford University
    - edX: "Practical Game Theory for Managers" by University of British Columbia
    - Khan Academy: "Game Theory" (free and awesome!)

- **Websites**:
    - Game Theory Explorer: An interactive tool to explore various game theory concepts (`http://gametheoryexplorer.org`)
    - Social Science Research Network: A platform for accessing game theory research papers (`https://www.ssrn.com`)
    - The Game Theory Society: Their website contains a wealth of information and resources (`https://gametheorysociety.org`)

So go forth, my friend, and embrace the power of game theory. It's a mind-bending journey that will equip you with the strategic savvy to navigate the complex game of life. Remember, play smart, not hard.

How Game Theory Applies to Everyday Life

Game theory may sound like some fancy academic shit that has zero relevance to your everyday life, but believe me, it's way more relatable than you think. In this section, we'll explore how game theory can be applied in various real-life situations, helping you make smart decisions and achieve your goals.

Negotiating for a Better Deal

Ever been in a situation where you had to negotiate with someone to get what you want? Maybe you were bargaining for a lower price on a used car, or maybe you were trying to convince your parents to let you stay out later. Well, guess what? Game theory has your back.

When it comes to negotiations, game theory provides a framework to analyze the strategies and outcomes of the involved parties. It helps you understand the importance of information, power dynamics, and the value of compromise. By thinking strategically and considering the other person's interests, you can maximize your chances of getting a better deal.

For example, let's say you're buying a car from a private seller. You know the maximum price you're willing to pay, but you also know that the seller wants to get as much money as possible. Applying game theory, you can identify potential bargaining chips, like pointing out any flaws in the car or mentioning alternative options you're considering. By strategically revealing information and understanding the seller's preferences, you increase your chances of reaching a mutually beneficial agreement.

Navigating Social Interactions

Social interactions can sometimes feel like a damn minefield. Whether you're dealing with cliques at school or office politics at work, game theory can help you navigate these complex dynamics with ease.

Game theory teaches us about the concept of equilibrium, where each individual's actions are in their best interest given the actions of others. Understanding this concept can help you anticipate how others might react in certain situations and adjust your own behavior accordingly.

Let's take a common scenario: deciding whether to cooperate or compete with your colleagues on a group project. Game theory shows us that cooperation can lead to better outcomes for everyone involved. However, it also highlights the risk of being taken advantage of if others choose to compete instead. By strategically assessing the dynamics within your group and finding the right balance between cooperation and competition, you increase your chances of achieving success while maintaining positive relationships.

Strategizing in Dating and Relationships

Yep, you read that right – game theory even applies to the ever-complicated world of dating and relationships. By understanding the principles of game theory, you can approach matters of the heart with a sharper mind and avoid getting caught up in toxic patterns.

Let's dive into a classic example: the dating "game" of playing hard to get. Game theory reveals that playing hard to get can sometimes work, as it increases your perceived value and creates a sense of desire in the other person. However, it's a delicate balance, as going too hard to get can lead to disinterest or frustration. By considering both your own desires and the desires of the other person, you can strategize your approach to build a healthy and fulfilling relationship.

Similarly, game theory helps you understand the concept of trust and the importance of cooperation in long-term relationships. By recognizing the benefits of trust and communication, you can work towards creating a mutually satisfying partnership.

Making Decisions in Uncertain Situations

Life is full of uncertainties, and often we're forced to make decisions without having all the information we desire. This is where game theory comes in handy.

Game theory provides tools to analyze decision-making under uncertainty. It helps you evaluate the potential outcomes and payoffs associated with different choices, even when the probabilities are unknown. By considering the potential strategies of others and the risks involved, you can make more informed decisions.

For example, let's say you're trying to decide whether to invest in a particular stock. Game theory allows you to consider various factors, such as market trends, competitor behavior, and investor sentiments. By analyzing the possible outcomes and payoffs from different scenarios, you can make a more rational investment decision.

Dealing with Conflict and Competition

Conflict and competition are part and parcel of life. Whether it's resolving disputes with friends or competing for a promotion at work, game theory offers insights on how to handle these situations effectively.

Game theory teaches us about strategies like the "tit-for-tat" approach, which involves reciprocating the actions of others. By employing this strategy, you can encourage cooperation and discourage selfish behavior. However, game theory also highlights the role of forgiveness and the importance of breaking negative cycles.

Consider a situation where you have a disagreement with a close friend. Applying game theory, you can assess the potential outcomes of different approaches, such as retaliating, compromising, or simply forgiving and moving forward. By strategically choosing your response, you increase the chances of resolving the conflict and preserving the relationship.

Resources and Further Reading

If you're interested in diving deeper into game theory and its applications, check out these resources:

- "Thinking Strategically" by Avinash K. Dixit and Barry J. Nalebuff: This book provides a comprehensive introduction to game theory, exploring its applications in various fields.

- "The Evolution of Cooperation" by Robert Axelrod: Axelrod delves into the study of cooperation, using game theory to understand why and how cooperation can emerge in certain situations.

- Online platforms like Coursera and Khan Academy offer courses on game theory taught by renowned professors. These courses provide interactive learning experiences and practical applications of game theory concepts.

Now that you've got a taste of how game theory can be applied to everyday life, it's time to level up your decision-making skills. Remember, by embracing game theory, you'll be playing smart, not hard, in all areas of your life.

Debunking Common Misconceptions about Game Theory

Game theory is a powerful tool for analyzing strategic interactions and decision-making in a wide range of fields. However, there are many misconceptions about game theory that can cloud our understanding and hinder its application. In this section, we will debunk some of these common misconceptions and clarify the true nature of game theory.

Misconception 1: Game theory is only about games

Contrary to its name, game theory is not solely about playing games. While it does have its roots in analyzing competitive situations, game theory's applications extend far beyond traditional games. In reality, game theory is a mathematical framework for studying strategic behavior and decision-making in any situation where the outcome depends on the choices of multiple actors. It has found practical applications in economics, politics, biology, psychology, and even computer science.

Misconception 2: Game theory only applies to competitive situations

Another common misconception is that game theory is only relevant in competitive scenarios. While game theory does have a strong foundation in analyzing competition, it is equally applicable in cooperative situations. Cooperative game theory focuses on studying the formation and stability of coalitions, negotiation strategies, and the distribution of resources in cooperative settings. Game theory is thus just as useful in understanding how cooperation emerges and is sustained, as it is in explaining competitive interactions.

Misconception 3: Game theory assumes rationality

One of the most persistent misconceptions about game theory is that it assumes all players are perfectly rational. While rationality is a fundamental concept in game theory, it does not imply that players always make perfect decisions. In reality, game theory acknowledges that individuals have limited information, cognitive biases, and emotions that can influence their decision-making. Behavioral game theory, a branch of game theory, incorporates these factors into its analysis, providing a more accurate representation of human behavior.

Misconception 4: Game theory is only applicable to humans

Game theory is not limited to analyzing interactions between humans. It has been successfully applied to understand the behavior of animals, bacteria, and even artificial intelligence agents. The principles of game theory apply as long as there are strategic choices and interdependencies between decision-makers. By studying how various species and systems make decisions, game theory can shed light on the evolution of strategies and behaviors across different domains.

Misconception 5: Game theory predicts outcomes with certainty

While game theory provides valuable insights into strategic interactions, it does not guarantee precise predictions of outcomes. Instead, it offers a framework for analyzing possible strategies and outcomes based on the assumptions and information available. The real world is often complex, with unpredictable variables and evolving dynamics. Game theory helps us understand the range of possible outcomes and anticipate how different strategies may play out, but it does not provide definitive predictions.

Misconception 6: Game theory promotes selfish behavior

Some mistakenly believe that game theory only encourages individuals to act in their own self-interest. While self-interest is indeed a significant driver of behavior in many situations, game theory also recognizes the importance of cooperation and trust. It studies how individuals can achieve better outcomes by cooperating and forming mutually beneficial relationships. In fact, game theory provides insights into how cooperation can emerge and be sustained, even in situations where purely self-interested behavior could lead to suboptimal outcomes.

Misconception 7: Game theory is too theoretical for practical use

There is a misconception that game theory is a purely theoretical construct with limited practical applications. However, game theory has been successfully applied to various real-world problems, ranging from economics and business strategy to political negotiations and cybersecurity. It provides valuable insights into how individuals and organizations make strategic decisions and helps identify optimal strategies in complex and uncertain environments.

Misconception 8: Game theory is outdated in the age of AI

As artificial intelligence (AI) continues to advance, some may question the relevance of game theory. However, game theory is more crucial than ever in understanding and shaping the interactions between humans and AI systems. AI algorithms are often designed to make decisions strategically, and game theory provides a framework to analyze and predict their behavior. Additionally, game theory is essential for studying the dynamics of multi-agent systems, where AI agents interact and make decisions in complex environments.

In conclusion, game theory is not just about games nor limited to competitive scenarios. It provides a powerful framework for understanding strategic interactions and decision-making in a wide range of fields. By debunking these common misconceptions, we can appreciate the true value and applicability of game theory in various domains. So, let's dive deeper into the fascinating world of game theory and discover how it can enhance our understanding of the complexities of decision-making in our everyday lives.

Key Takeaways:

- Game theory extends beyond traditional games and is applicable in various fields, including economics, politics, biology, and computer science.

- It analyzes both competitive and cooperative interactions, providing insights into how cooperation emerges and is sustained.

- Game theory acknowledges that players may have limited information, cognitive biases, and emotions that influence their decision-making.

- It is not confined to human decision-making but also encompasses the behavior of animals, bacteria, and AI agents.
- While game theory helps analyze possible strategies and outcomes, it does not guarantee precise predictions due to the complexity of real-world dynamics.
- Game theory recognizes the importance of cooperation and offers insights into how it can lead to better outcomes.
- It has practical applications in economics, business, politics, and cybersecurity, among others.
- Game theory remains relevant in the age of AI, as it helps analyze the behavior of AI systems and understand multi-agent interactions.

The Role of Rationality in Game Theory

In game theory, rationality is a fundamental assumption that underlies the behavior of players. It is based on the idea that individuals act in a self-interested manner, with the goal of maximizing their own outcomes. Rationality provides a framework for analyzing strategic interactions and predicting the behavior of players.

The Rational Choice Model

The rational choice model forms the basis of rationality in game theory. According to this model, individuals are rational actors who carefully consider the available information, options, and consequences before making decisions. They aim to maximize their own utility, or satisfaction, based on their personal preferences.

In strategic games, rationality implies that players will choose strategies that lead to the best outcome for themselves, taking into account their beliefs about the actions of other players. This assumption enables the prediction of player behavior and the identification of equilibrium points in a game.

Cognitive Biases and Decision-Making

While rationality assumes that individuals make decisions based on careful analysis, research in psychology has shown that people often deviate from rational behavior due to cognitive biases. These biases are systematic errors in decision-making that arise from mental shortcuts and heuristics.

One example of a cognitive bias is the confirmation bias, which leads individuals to seek out information that confirms their preexisting beliefs and ignore contradictory evidence. Another bias is the framing effect, where individuals react differently to the same information depending on how it is presented to them.

Understanding cognitive biases is crucial in game theory, as it helps explain why players might deviate from rational behavior and make decisions that are suboptimal for themselves. By incorporating these biases into game models, researchers can gain a more realistic understanding of strategic interactions.

The Role of Emotions in Decision-Making

Emotions play a significant role in decision-making, and game theory acknowledges their influence on strategic behavior. While rationality assumes that individuals are purely rational and devoid of emotions, research has shown that emotions can greatly impact decision-making processes.

Emotional states such as anger, fear, and happiness can alter a player's perception of the game and their willingness to take risks. For example, in a competition, a player who is angry may become more aggressive and more willing to engage in confrontational strategies. Conversely, a player in a state of fear may become more cautious and conservative in their decision-making.

Accounting for the role of emotions in game theory can lead to a more nuanced understanding of strategic interactions and the motivations behind certain behaviors.

Analyzing Risk and Uncertainty

Risk and uncertainty are inherent in game theory, as players often make decisions under conditions of incomplete information. Rational decision-making involves assessing the potential risks and uncertainties associated with different strategies and taking them into account when making choices.

One way to analyze risk in game theory is through the concept of expected utility. Expected utility theory assumes that players assign utilities to different outcomes and make decisions based on the expected value of these utilities. This approach allows for a quantitative assessment of the risk associated with different strategies.

Uncertainty, on the other hand, arises when players do not have enough information to assign probabilities to various outcomes. In such cases, players may rely on probabilistic reasoning or adopt more cautious strategies to mitigate the potential negative consequences.

Prospect Theory: An Alternative to Rational Choice

While the rational choice model provides a useful framework for analyzing decision-making, it does not fully capture the complexities of human behavior. Prospect theory, developed by Daniel Kahneman and Amos Tversky, offers an alternative approach that accounts for biases and deviations from rationality.

Prospect theory is based on the idea that individuals evaluate outcomes relative to a reference point, rather than in absolute terms. It also considers the notion of diminishing sensitivity, where the impact of gains and losses diminishes as the magnitude increases.

This alternative model of decision-making provides a more accurate representation of human behavior in situations involving risk and uncertainty. By incorporating prospect theory into game theory, researchers can better understand and predict decision-making in complex strategic interactions.

Bounded Rationality: Making the Best of Limited Information

Bounded rationality is a concept that recognizes the cognitive limitations and information constraints faced by individuals in decision-making processes. It posits that individuals strive to make rational decisions but are limited by factors such as time, cognitive resources, and the complexity of the decision problem.

In game theory, bounded rationality acknowledges that players may not have complete knowledge of the game structure, their opponents' preferences, or the potential consequences of their actions. Despite these limitations, players aim to make the best decisions based on the available information and their cognitive capabilities.

Herbert Simon, a Nobel laureate, introduced the idea of "satisficing" as a form of bounded rationality. Rather than searching for the best possible solution, individuals satisfice by selecting the first option that meets a satisfactory level of utility. This approach helps overcome information constraints and simplifies decision-making in complex strategic contexts.

Decision-Making in Competitive and Cooperative Games

The role of rationality in decision-making varies in competitive and cooperative games. In competitive games, rationality often leads players to adopt aggressive and self-interested strategies as they aim to outperform their opponents and maximize their own outcomes.

In cooperative games, however, rationality can also lead to cooperative behavior. Players may recognize that cooperating and forming coalitions can lead to higher payoffs for all involved. The rational choice model allows for the analysis of cooperative strategies and the identification of stable coalitions.

The role of rationality in decision-making in both competitive and cooperative games depends on various factors, including the game structure, the preferences of players, and the incentives at play.

Real-Life Applications of Decision-Making in Game Theory

Decision-making in game theory has numerous real-life applications across different fields. In economics, it helps analyze market competition, pricing strategies, and auction formats. In politics and international relations, it sheds light on negotiation processes, voting behavior, and conflict resolution.

Game theory also finds applications in social sciences, such as sociology and psychology, where it helps understand social interactions and collective decision-making. It is used in biology and ecology to study animal behavior and evolutionary dynamics.

Furthermore, decision-making in game theory is relevant in everyday life. Situations like buying a car, negotiating a salary, or even playing sports involve strategic interactions where rational decision-making can affect outcomes.

Understanding the role of rationality in game theory allows individuals to make better choices, anticipate the behavior of others, and navigate complex social and economic situations.

Exercises

1. Consider a competitive game between two companies, A and B, in the smartphone market. Analyze the role of rationality in their decision-making processes. How might cognitive biases impact their strategies? 2. Think of a real-life situation where emotions could influence the outcome of a game. Explain how the players' emotional states might affect their decisions. 3. Research a historical event or conflict and apply game theory to analyze the decision-making processes of the involved parties. Consider the role of rationality, risk assessment, and strategic thinking. 4. Imagine a cooperative game where a group of friends needs to plan a vacation together. Apply the concepts of rationality and bounded rationality to analyze their decision-making and the formation of coalitions. 5. Investigate a contemporary issue, such as climate change or healthcare policy, and discuss how game theory can be used to understand the decision-making processes of different stakeholders. Consider the influences of rationality and bounded rationality.

Recommended Resources

Books: - "Thinking, Fast and Slow" by Daniel Kahneman - "The Strategy of Conflict" by Thomas C. Schelling - "Game Theory: A Very Short Introduction" by Ken Binmore

Online Courses: - "Game Theory" on Coursera, offered by Stanford University - "An Introduction to Game Theory" on edX, offered by the University of Tokyo

Websites: - Game Theory Society (www.gametheorysociety.org) - Stanford Encyclopedia of Philosophy - Game Theory (plato.stanford.edu/entries/game-theory)

Podcasts: - "Game Theory with Sam Vecenie" hosted by Sam Vecenie - "The Knowledge Project with Shane Parrish" - Episode on mental models and decision-making

Further Reading

1. Camerer, C. F. (2003). Behavioral Game Theory: Experiments in Strategic Interaction. Princeton University Press. 2. Osborne, M. J., & Rubinstein, A. (1994). A Course in Game Theory. MIT Press. 3. Dixit, A. K., & Nalebuff, B.J. (2008). The Art of Strategy: A Game Theorist's Guide to Success in Business and Life. WW Norton & Company. 4. Fudenberg, D., & Tirole, J. (1991). Game Theory. MIT Press. 5. Schelling, T.C. (1960). The Strategy of Conflict. Harvard University Press.

Ethics and Game Theory: Playing Fair or Cheating?

In the game of life, fairness is often seen as a guiding principle. We are taught from a young age to play by the rules, to treat others as we would like to be treated, and to value honesty and integrity. But when it comes to game theory, things can get a little more complicated. Game theory is all about strategic decision-making in situations where the outcome depends not only on our own choices but also on the choices of others.

So, what role do ethics play in game theory? Can we still play fair when the stakes are high and the competition is fierce? Or does game theory push us towards cheating and deception? Let's dive in and explore the ethical considerations in game theory.

Understanding Ethics and Game Theory

Ethics, in the context of game theory, refers to the moral principles that guide our behavior and decisions. It involves evaluating what is right or wrong, fair or unfair, just or unjust, in the strategic interactions we encounter. Game theory helps us understand how individuals or groups make decisions and act in situations where their interests may conflict.

At its core, game theory provides a framework for analyzing the strategic interactions between different players. It allows us to study the possible outcomes, strategies, and payoffs in games. However, it does not prescribe any particular ethical rules or values. Instead, it offers a lens through which we can examine the ethical implications of the choices we make.

Playing Fair: Cooperation and Trust

In many games, cooperation and trust can lead to mutually beneficial outcomes. The classic example of cooperation is the Prisoner's Dilemma, where two individuals have to decide whether to cooperate or betray each other. If both players choose to cooperate, they receive a moderate payoff. However, if one player betrays the other while the other cooperates, the betrayer receives a higher payoff, and the cooperating player receives a lower payoff. If both players betray each other, they both receive a lower payoff.

In the context of ethics, playing fair means choosing to cooperate rather than betraying others in pursuit of personal gain. It involves considering the long-term consequences, building trust, and promoting cooperation for the greater good. Playing fair can create a more harmonious and cooperative society, leading to better outcomes for all.

The Temptation of Cheating

While playing fair may seem like the morally superior choice, game theory also reveals the temptation to cheat. Cheating can provide short-term gains and put an individual at an advantage over others. In games with incomplete or imperfect information, players may attempt to deceive others to gain a strategic advantage.

However, cheating can have negative consequences not only for the players involved but also for the overall stability of the game. If trust is eroded, cooperation may break down, and everyone suffers in the long run. In some cases, the cost of cheating can outweigh the benefits, leading to a suboptimal outcome for all players involved.

Ethical Dilemmas in Game Theory

Game theory presents various ethical dilemmas that require careful consideration. One such dilemma is the tension between competition and fairness. While competition is often valued in game theory, it can sometimes lead to unfair outcomes, especially when there is a significant power imbalance between players.

Another ethical dilemma arises when considering the treatment of individuals versus the pursuit of overall societal well-being. Game theory often focuses on the collective outcomes and maximizing overall utility. However, this can sometimes neglect the rights and interests of individual players. Striking a balance between individual welfare and collective well-being is crucial when making ethical choices in strategic interactions.

Preserving Ethics in Game Theory

So, how can we ensure that ethics are preserved in game theory? One approach is to introduce mechanisms that promote fairness and cooperation. For example, in repeated games, where interactions occur multiple times, tit-for-tat strategies can encourage cooperation by reciprocating the other player's previous move. This promotes trust and discourages cheating.

Transparency and communication can also help preserve ethics in strategic interactions. When players have complete and accurate information, they can make more informed decisions and reduce the temptation to cheat. Open dialogue allows for negotiation, compromise, and building trust among players.

Moreover, considering the long-term consequences and impact on society as a whole can guide ethical decision-making in game theory. By prioritizing fairness and cooperation, we can create a more equitable and sustainable society.

Real-World Examples

The ethical implications of game theory can be observed in various real-world scenarios. For instance, in business, companies must balance competition with ethical business practices. Unfair market practices, such as price-fixing, can harm consumers and lead to legal consequences. On the other hand, ethical business practices can improve brand reputation and long-term success.

In politics, ethical considerations play a crucial role in decision-making. Politicians must balance their self-interest with the well-being of their constituents. Transparency, accountability, and fairness in policy-making can foster trust and promote the common good.

Conclusion

Ethics and game theory are intertwined. While game theory provides a framework for strategic decision-making, it is up to us to apply ethical principles and values in our choices. Playing fair and promoting cooperation can lead to better outcomes for all players involved. Cheating may provide short-term gains, but it often comes at the cost of trust and long-term stability. By considering the ethical implications of game theory, we can create a more just and sustainable society. Remember, in the game of life, it's not just about winning—it's about how we play the game.

Game Theory in a Digital Age: Online Gaming and Virtual Economies

With the rapid advancement of technology and the rise of online gaming, the field of game theory has found new avenues for exploration. Today, we dive into the fascinating world of online gaming and virtual economies, and examine how game theory can help us understand and navigate these digital landscapes.

The Rise of Online Gaming

Online gaming has taken the world by storm, attracting millions of players from all walks of life. The popularity of games such as World of Warcraft, Fortnite, and League of Legends has led to the emergence of massive virtual communities, where players interact, compete, and collaborate in a rich gaming environment.

Virtual Economies: The Game within the Game

One of the most intriguing aspects of online gaming is the existence of virtual economies. These economies, comprised of in-game currencies, virtual goods, and trading systems, mimic real-world markets in many ways. Players can buy and sell items, engage in virtual trading, and accumulate wealth within the confines of the game.

In virtual economies, game theory provides valuable insights into the behavior of players and the dynamics of the market. Traditional economic theories struggle to explain the unique characteristics of these virtual ecosystems, making game theory a natural fit for understanding their intricacies.

Strategic Decision-Making in Virtual Economies

Game theory teaches us that rational decision-making is essential in any strategic interaction, and virtual economies are no exception. Players must make informed choices about when to buy or sell, how to negotiate prices, and when to invest resources.

One fascinating concept in virtual economies is the notion of *scarcity*. Just like in the real world, the supply and demand of virtual goods can greatly influence their value. Understanding the impact of scarcity on pricing and trading dynamics requires a game-theoretic perspective.

Market Manipulation and Exploitation

Virtual economies are not immune to manipulation and exploitation. Just as players can form alliances and cooperate, they can also engage in deceitful practices to gain an unfair advantage. This can include manipulating prices, cornering markets, or using illegal third-party tools to accumulate wealth.

Game theory provides a framework for identifying and combating such behaviors. By analyzing strategic interactions and equilibria, we can devise mechanisms to discourage exploitative practices and ensure a fair and enjoyable gaming experience for all.

Regulation and Governance in Virtual Economies

The importance of regulation and governance in virtual economies cannot be overstated. Just like real-world markets, virtual economies require rules and mechanisms to curb fraud, prevent market crashes, and protect the rights of participants.

Game theory offers strategies for designing effective regulatory frameworks. Mechanisms such as taxation, trade restrictions, and anti-fraud measures can be implemented to maintain a stable, sustainable virtual economy.

The Impact of Virtual Economies on the Real World

Virtual economies are not confined to the gaming realm; they often have real-world consequences. The sale of virtual items and currencies for real money, known as *real-money trading*, has created a multi-million-dollar industry. Additionally, virtual economies can impact player behavior, social dynamics, and even influence the design of future games.

Understanding the relationship between virtual and real-world economies is a complex challenge, but game theory provides a toolset to analyze these interactions and their implications for both worlds.

Case Study: EVE Online

A prime example of the intricate nature of virtual economies is the massively multiplayer online game EVE Online. In this game, players participate in a ruthless universe of space exploration, warfare, and economic maneuvering.

EVE Online's virtual economy is known for its complexity and the intricate interplay of supply and demand. The game features a player-driven market where goods are produced, traded, and fought over. The decisions and strategies adopted by players can significantly impact the game's economy, making it a captivating case study for game theorists.

Challenges and Future Directions

As virtual economies continue to evolve and expand, new challenges and opportunities emerge. The integration of blockchain technology, the rise of non-fungible tokens (NFTs), and the advent of decentralized finance (DeFi) present exciting developments with profound implications for online gaming and virtual economies.

Game theory will undoubtedly play a crucial role in understanding and addressing these challenges. By embracing the dynamic nature of virtual economies and leveraging the power of strategic thinking, we can pave the way for a more immersive, fair, and economically vibrant digital gaming landscape.

As we navigate the ever-changing realm of online gaming and virtual economies, game theory provides invaluable insights into the strategic interactions that define these digital landscapes. Understanding the behavior of players, the dynamics of virtual markets, and the societal implications of these systems allows us to make smarter decisions and shape the future of online gaming.

Gaining an Advantage: Strategies and Tactics

In the world of game theory, gaining an advantage over your opponents is crucial. Whether you're playing a simple game of rock-paper-scissors or making strategic decisions in complex business negotiations, understanding the strategies and tactics that can give you the upper hand is essential. In this section, we will explore various techniques that can help you gain an advantage and increase your chances of success.

Strategies for Optimal Decision-Making

To gain an advantage in a game, you need to make optimal decisions. This means choosing the action that will maximize your payoff or minimize your loss. Rational decision-making theory provides a framework for analyzing and making such decisions.

One commonly used model is the rational choice model, which assumes that individuals act rationally and aim to maximize their utility. However, in reality, decision-making is often influenced by cognitive biases and emotions. Understanding these biases and emotions can help you make better decisions.

Cognitive biases, such as the anchoring bias or the availability bias, can cloud your judgment and lead to suboptimal decisions. By being aware of these biases, you can actively counteract them and make more rational choices.

Emotions also play a significant role in decision-making. Fear, greed, and overconfidence can all lead to irrational decisions. By being aware of your emotions and taking a step back to analyze the situation objectively, you can make better-informed choices.

In addition to cognitive biases and emotions, analyzing risk and uncertainty is vital when making decisions. Prospect theory, an alternative to the rational choice model, takes into account the way individuals perceive and evaluate risks. It suggests that people are more sensitive to losses than gains, which can influence decision-making. By understanding prospect theory, you can assess risk effectively and make decisions that maximize your gains.

Tactics for Outmaneuvering Your Opponents

While having a solid strategy is crucial, employing effective tactics can also give you an advantage. Let's explore some tactics that can help you outmaneuver your opponents in a game.

1. The first tactic is bluffing. Bluffing involves presenting false information to deceive your opponents. This tactic is commonly used in poker, where players try to make their opponents believe they have a better hand than they actually do. Bluffing can force your opponents to make suboptimal decisions or fold when they have a strong hand, giving you an advantage.

2. Another tactic is the use of mixed strategies. Instead of always choosing a fixed action, a player can adopt a mixed strategy by randomizing their choices. This unpredictability makes it challenging for opponents to exploit any patterns in your behavior. Mixed strategies are especially effective in games like the prisoner's dilemma.

3. Timing is everything in strategic games. By carefully selecting when to make your moves, you can gain an advantage over your opponents. This tactic is crucial in sequential games, where each player's move depends on the previous player's action. By analyzing the game tree and employing backward induction, you can identify strategic timing opportunities and make decisions accordingly.

4. Information asymmetry can be exploited to gain an advantage. If you have more information than your opponents, you can make more informed decisions. For example, in negotiations, gathering information about the other party's preferences and constraints can help you structure your offers strategically.

5. Building alliances and forming coalitions can also be effective tactics. By collaborating with like-minded individuals, you can pool resources, share information, and achieve mutually beneficial outcomes. This tactic is often observed in cooperative games, business partnerships, and even political alliances.

6. Finally, analyzing your opponent's behavior and adapting your strategies accordingly is a crucial tactic. This involves studying their patterns, strengths, and weaknesses and adjusting your gameplay to exploit their vulnerabilities. For example, if your opponent tends to be overly aggressive, you can adopt a more defensive strategy to counter their moves.

Putting Strategies and Tactics into Practice

To truly understand and apply strategies and tactics, it is important to explore real-world applications. Let's consider a few examples where gaining an advantage through strategic decision-making is critical.
 1. Business Negotiations: In negotiations, understanding your opponent's motivations, values, and constraints can allow you to make offers that provide them with value while maximizing your own gains. By employing tactics such as information gathering, timing your offers strategically, and using creative problem-solving, you can secure the best possible outcomes.
 2. Online Gaming: In the world of online gaming, players often employ tactics such as bluffing, timing their moves effectively, and analyzing their opponents' behavior to gain an advantage. By understanding the game mechanics and employing common strategies, skilled players can outwit their opponents and achieve victory.
 3. Financial Markets: Traders and investors in financial markets constantly analyze market trends, gather information, and adapt their strategies to gain an advantage. Employing tactics such as technical analysis, risk management, and staying updated with market news can help investors maximize their profits and minimize losses.

Conclusion

Gaining an advantage in games, whether they are played for fun or in real-life scenarios, requires both strategic thinking and tactical maneuvering. By understanding the principles of optimal decision-making, being aware of cognitive biases and emotions, and employing effective tactics, you can increase your chances of success. Remember, gaining an advantage is not about cheating or unfair play, but rather about using your knowledge and skills to outmaneuver your opponents and achieve optimal outcomes. So, play smart, not hard, and may the odds be ever in your favor!

Exercises

1. In a game of rock-paper-scissors, what tactics can you employ to gain an advantage over your opponents? Consider both strategic thinking and tactical maneuvering.

2. In a negotiation, how can you gather information about the other party's preferences and constraints? How can this information be used strategically to gain an advantage?

3. Choose a real-life scenario where gaining an advantage through strategic decision-making is critical. Analyze the tactics that can be employed in that scenario and discuss their potential outcomes.

4. Consider a game where you have limited information about your opponents' strategies. How can you use mixed strategies to gain an advantage? Provide a step-by-step analysis.

5. Imagine you are playing a sequential game with multiple players. How can you determine the optimal timing of your moves to gain an advantage? Provide a detailed explanation.

Additional Resources

1. Owen, G., & Shapley, L. (2009). *Learning to cooperate in repeated games: Stochastic and deterministic behavior*. Princeton University Press.

2. Dixit, A., & Nalebuff, B. J. (2009). *The art of strategy: A game theorist's guide to success in business and life*. WW Norton & Company.

3. Binmore, K. (2007). *Playing for real: A text on game theory*. Oxford University Press.

4. Krueger, A. O. (2018). *Doing deals: The art of business transaction management*. Harvard Business Press.

5. Kahneman, D., Tversky, A., & Thaler, R. (1991). *The endowment effect, loss aversion, and status quo bias*. Journal of Economic Perspectives, 5(1), 193-206.

Remember, gaining an advantage is not about cheating or unfair play, but rather about using your knowledge and skills to outmaneuver your opponents and achieve optimal outcomes. So, play smart, not hard, and may the odds be ever in your favor!

The Future of Game Theory: AI and Machine Learning

In recent years, the world has witnessed remarkable advancements in artificial intelligence (AI) and machine learning. These technologies have permeated almost every aspect of our lives, revolutionizing industries such as healthcare, finance, and transportation. Unsurprisingly, game theory has also been profoundly impacted by these developments. In this section, we will explore the future of game theory in the era of AI and machine learning, examining the exciting possibilities and potential challenges that lie ahead.

AI and the Enhancing of Strategic Decision-Making

One of the most significant contributions of AI to game theory is its ability to enhance strategic decision-making. Traditional game theory assumes that players are perfectly rational and have complete information. However, in reality, players often face uncertainty and are limited by their cognitive abilities.

Machine learning algorithms can learn from vast amounts of data and uncover patterns that human players might overlook. By analyzing the strategies and behaviors of past players, AI can identify optimal moves and make more informed decisions. This has wide-ranging implications for games involving complex strategic interactions, such as chess, poker, and even economic markets.

Consider an example of an AI system playing poker. Through reinforcement learning, the AI can learn the optimal strategy for each stage of the game by playing against itself or analyzing historical data. This enables the AI to make better decisions, bluff effectively, and adapt to its opponents' tactics. Consequently, AI is pushing the boundaries of what is considered "optimal" play in games, challenging human players to step up their game.

Game Theory and Multi-Agent Reinforcement Learning

Another exciting development in the future of game theory is the application of multi-agent reinforcement learning (MARL). MARL involves training multiple AI agents to interact with each other, similar to how humans interact in strategic games. This approach allows researchers to study complex interactions and strategies between intelligent agents.

MARL has already shown tremendous potential in solving complex games. For instance, researchers at OpenAI developed AlphaZero, a neural network-based system that achieved superhuman performance in chess, shogi, and Go by playing against itself. The AI agents in AlphaZero use reinforcement learning to continuously improve their strategies through self-play, resulting in unprecedented levels of mastery.

This advancement has significant implications beyond games. MARL could potentially be applied to real-world situations where multiple decision-makers interact, such as negotiations, auctions, and traffic flow. By understanding the dynamics and strategies of intelligent agents, game theory can provide valuable insights into optimal decision-making in these settings.

Ethical Considerations and Fairness in AI-based Game Theory

While the integration of AI and machine learning into game theory offers exciting possibilities, we must not overlook the ethical considerations that arise. One area of concern is ensuring fairness and preventing AI from exploiting loopholes or engaging in unethical behavior.

In game theory, fairness often plays a crucial role in decision-making. When AI agents are involved, the question of fairness becomes more complex. How do we ensure that AI agents follow ethical principles, even when there may be incentives to deviate? Can AI learn to play cooperatively and fairly, or will it exploit weaknesses in the game?

Moreover, the emergence of AI-powered decision-making systems raises questions about accountability and transparency. If AI agents make decisions that impact individuals or society, who bears the responsibility? How can we ensure that AI decisions are explainable and interpretable by humans? These questions require careful consideration to ensure the responsible and ethical use of AI in game theory.

The Collaborative Future: Human-AI Interaction

In the future, game theory will increasingly involve collaborative efforts between humans and AI. This collaboration has the potential to unlock new levels of strategic thinking and decision-making.

AI systems can serve as powerful tools to assist human players, offering insights, analyzing complex scenarios, and suggesting optimal strategies. Humans, on the other hand, can provide the AI with a nuanced understanding of the game, creative thinking, and the ability to adapt to changing circumstances.

Consider an AI chess coach that assists a human player. The AI can analyze vast amounts of chess data, study the player's weaknesses, and suggest training strategies. The human player, in turn, can contribute their intuition, insight, and emotional awareness to refine the AI's recommendations.

Such collaborations between humans and AI can also lead to the development of new hybrid game-playing entities that combine the best qualities of both. These entities could participate in tournaments and competitions, challenging existing game theory paradigms.

Conclusion

The future of game theory lies at the intersection of AI and machine learning. These technologies have the potential to enhance strategic decision-making, enable the development of sophisticated game-playing algorithms, and foster collaborative efforts between humans and AI. However, the integration of AI into game theory also raises ethical considerations that must be addressed.

As game theory continues to evolve, embracing AI and machine learning will shape the future landscape of strategic decision-making. The possibilities are vast, and we are only beginning to scratch the surface of what can be achieved. It is an exciting time for game theory, offering new opportunities for innovation, collaboration, and understanding in the ever-evolving world of games.

Decision-Making in Game Theory

Rationality and Decision-Making

The Rational Choice Model

In the world of game theory, decision-making is a key concept that underlies many strategic interactions. At the heart of decision-making lies the rational choice model, which assumes that individuals make choices based on a process of deliberation, considering all available information and weighing the costs and benefits of each option.

The rational choice model, also known as the rational actor model, is built on the assumption that individuals are rational beings who maximize their self-interest and aim to achieve the best possible outcome. This model forms the foundation of classical economics and has been applied to various fields, including psychology, sociology, political science, and philosophy.

According to the rational choice model, individuals are motivated by their preferences and aim to make choices that satisfy those preferences. These preferences can vary from person to person and may include factors such as personal values, desires, goals, and beliefs. The rational choice model assumes that individuals are aware of their preferences and have a clear understanding of the available options.

In order to make a rational choice, individuals must also have sufficient information about the consequences of their choices. This information includes knowledge about the potential outcomes, the probability of each outcome occurring, and the associated costs and benefits. The rational choice model assumes that individuals have access to this information and can accurately assess it.

To illustrate the rational choice model, let's consider a simple example. Suppose you are a college student deciding whether to attend a party or stay home and study for an upcoming exam. You have a preference for having fun and socializing, but you also value academic success. In this case, the rational choice model suggests that you would weigh the potential benefits of attending the party, such as enjoyment and social connection, against the potential costs, such as reduced study time and a lower exam score. Based on this assessment, you would then make a rational decision that aligns with your preferences and maximizes your self-interest.

However, it is important to note that the rational choice model has its limitations and does not capture the full complexity of human decision-making. In reality, individuals may not always have complete information, may have cognitive biases that affect their judgment, or may be influenced by emotions or external pressures. These deviations from rationality are known as bounded rationality and can lead to suboptimal decision-making.

Despite its limitations, the rational choice model remains a powerful tool for analyzing decision-making and understanding strategic interactions. It provides a framework for predicting and explaining behavior in various contexts, such as economics, politics, and social sciences. By understanding how individuals make rational choices, researchers and policymakers can develop strategies to influence behavior, design incentives, and promote desirable outcomes.

To further explore the rational choice model, let's consider a real-world example. Imagine a scenario where two companies are deciding whether to collaborate on a project or pursue individual strategies. Each company must weigh the potential benefits of collaboration, such as increased resources and expertise, against the potential costs, such as sharing profits and compromising autonomy. By analyzing the decision-making process using the rational choice model, we can assess the likely outcomes and understand the incentives that drive the behavior of the companies.

In conclusion, the rational choice model provides a framework for understanding decision-making in game theory. It assumes that individuals are rational actors who aim to maximize their self-interest and make choices based on a careful assessment of available information and preferences. While the model has its limitations, it remains a valuable tool for analyzing behavior in various fields. By studying the rational choice model, we can gain insights into human decision-making and develop strategies to influence behavior and achieve desirable outcomes.

Cognitive Biases and Decision-Making

In the fascinating world of game theory, understanding decision-making is pivotal to success. It's not just about weighing the pros and cons, my friend, it's about delving into the intricate workings of the human mind. Even though we like to think of ourselves as rational beings, our decisions are often influenced by a myriad of cognitive biases that can lead us astray. In this section, we will unravel the mysteries of these biases and explore how they shape our decision-making process in game theory.

Biases Galore

Let's kick things off by diving into some of the most common cognitive biases that impact our decision-making. Prepare yourself, because these biases are sneaky little devils.

1. Confirmation Bias: Ah, the beauty of cherry-picking information that confirms our preconceived notions. We tend to seek out evidence that supports our beliefs while ignoring or downplaying contradictory evidence. In game theory, this can lead to flawed decision-making if we only seek out information that supports our favored strategy.

2. Anchoring Bias: How easily we get swayed by the first piece of information we encounter. This bias occurs when we rely too heavily on the initial information (the anchor) when making decisions. In game theory, this can manifest when players place too much importance on the first offer made in a negotiation or the starting bid in an auction.

3. Availability Heuristic: Our minds love shortcuts, and the availability heuristic is one such shortcut. It occurs when we make judgments based on the ease with which examples or instances come to mind. In game theory, this can lead to biased decision-making if we overestimate the likelihood of certain outcomes based on vivid or recent examples.

4. Overconfidence Bias: Ah, the illusion of superiority. We often have more faith in our abilities and judgments than is warranted. In game theory, this bias can lead us to overestimate our chances of success or underestimate the skills of our opponents, leading to disastrous outcomes.

5. Framing Effect: How the presentation of information can drastically change our decision-making. The framing effect occurs when we make different choices based on how options are presented, even if the content is the same. In game theory, this bias can be exploited through strategic framing to influence our opponents' decisions.

6. Loss Aversion: The fear of losing is a powerful motivator. Loss aversion refers to our tendency to prefer avoiding losses over acquiring equivalent gains. In game theory, this bias can lead to risk-averse behavior, where players avoid risky strategies that could potentially lead to losses.

Navigating the Bias Minefield

Now that we understand some of the cognitive biases that can trip us up, how can we navigate this treacherous minefield and make better decisions in game theory? Fear not, my friend, for I have some tricks up my sleeve.

1. Awareness is Key: The first step in overcoming biases is to be aware of them. By recognizing the biases that are at play, we can take steps to counteract their influence on our decision-making. Stay vigilant, my friend!

2. Diverse Perspectives: Seek out diverse opinions and perspectives, my friend. By considering different viewpoints, we can challenge our own biases and make more informed decisions. In game theory, this can be valuable when forming alliances or negotiating with others.

3. Slow Down: Our minds love shortcuts and snap judgments, but sometimes it's important to slow down and deliberate. Take your time, my friend, and carefully consider the consequences of your decisions in game theory. Patience truly is a virtue.

4. Test and Learn: Embrace experimentation, my friend. Test different strategies, learn from the outcomes, and refine your approach. By embracing a mindset of continuous learning, we can adapt and improve our decision-making skills in game theory.

5. Seek Feedback: Don't be afraid to seek feedback from others, my friend. Peers, mentors, and experts can provide valuable insights and help mitigate the impact of cognitive biases on our decision-making. Embrace constructive criticism and learn from the wisdom of others.

Breaking the Mold

Before we wrap up this section, let's dive into a contemporary example that showcases the impact of cognitive biases on decision-making in game theory.

Imagine a scenario where two companies are competing for a contract. Both companies have similar qualifications and expertise. However, Company A has a charismatic CEO who is a persuasive speaker, while Company B has a CEO known for being a meticulous planner.

Due to the availability heuristic, the client's decision-makers are more likely to remember Company A and the charismatic CEO. Even though both companies are equally qualified, the framing effect could lead the decision-makers to perceive charisma as a more desirable trait for the project. This bias might lead the client to choose Company A over Company B, solely based on the CEO's presentation skills.

In this example, the biases of availability and framing can heavily influence the decision-making process and potentially affect the outcome of the game between the two companies.

Further Reading

To further expand your understanding of cognitive biases and decision-making in game theory, I highly recommend the following resources:

1. "Thinking, Fast and Slow" by Daniel Kahneman

2. "Nudge: Improving Decisions About Health, Wealth, and Happiness" by Richard H. Thaler and Cass R. Sunstein

3. "Predictably Irrational: The Hidden Forces That Shape Our Decisions" by Dan Ariely

These books delve deep into the fascinating world of cognitive biases and offer valuable insights into how our decision-making processes are influenced.

Exercises

To test your newfound knowledge, here are a few exercises for you, my friend:

1. Identify a cognitive bias that you have personally experienced in your own decision-making. Reflect on the situation and analyze how the bias impacted your choices. How might you approach the situation differently in the future?

2. Think of a real-world game or strategic situation where cognitive biases could come into play. Describe how these biases could influence the decisions made by the players involved.

3. Conduct a small experiment with your friends or classmates to observe the impact of framing effects on decision-making. Present the same scenario, but frame it in different ways to see if the choices made by participants are influenced by the framing.

Get out there, my friend, and explore the fascinating world of cognitive biases and decision-making in game theory. Armed with the knowledge of these biases, you're well on your way to becoming a strategic mastermind.

Remember, in the game of life, it's not just about playing hard; it's about playing smart.

The Role of Emotions in Decision-Making

Emotions play a vital role in decision-making. While we often think of decision-making as a rational process based solely on logic, the truth is that our emotions heavily influence the choices we make. In this section, we will explore the role of emotions in decision-making, the impact they have on our choices, and how understanding this relationship can help us make better decisions.

The Influence of Emotions

At its core, decision-making is a complex interplay between our rational thinking and our emotions. Research has shown that emotions can significantly impact our judgment and decision-making processes, sometimes even overriding logic.

One influential theory in this area is the somatic marker hypothesis proposed by neuroscientist Antonio Damasio. According to this theory, emotions serve as mental shortcuts or markers that guide our decision-making. When faced with a decision, our emotions provide us with a quick evaluation of possible outcomes based on past experiences, helping us determine the best course of action.

For example, imagine you are considering whether to invest in a particular stock. Your rational thinking might involve analyzing financial data and market trends. However, your emotions may come into play when you recall a previous investment that resulted in a significant loss. This emotional memory may create a somatic marker, cautioning you against making a similar risky investment.

Emotions as Decision Heuristics

Emotions can also act as decision heuristics, mental shortcuts that simplify the decision-making process. Rather than weighing all available information, we often rely on our emotions to make quick decisions.

For instance, consider a situation where you are deciding between two job offers. One offer promises a higher salary, while the other offers a more flexible work schedule. Your emotions might lead you to choose the job that brings you joy and fulfillment, even if it means sacrificing some financial gain.

However, it is essential to recognize that emotions can sometimes lead to irrational decision-making. We may be swayed by fear, anger, or excitement, causing us to make choices that are not in our best interest. It is crucial to balance our emotions with rational thinking to make well-informed decisions.

The Role of Emotional Intelligence

Emotional intelligence is the ability to understand and manage our emotions effectively. Developing emotional intelligence can help us make better decisions by allowing us to recognize and regulate our emotions during the decision-making process.

By cultivating emotional intelligence, we can become more aware of the emotions that influence our choices. We can then take a step back, assess the situation objectively, and evaluate whether our emotions are clouding our judgment.

Additionally, emotional intelligence helps us understand the emotions of others, enabling us to navigate social dynamics effectively. In situations involving negotiations or collaborations, recognizing and empathizing with the emotions of others can lead to better outcomes.

Harnessing Emotions for Better Decision-Making

Rather than trying to suppress or ignore our emotions, we can harness them to our advantage in decision-making. Here are some strategies to help us make better decisions by incorporating our emotions:

1. Pause and Reflect: When faced with a decision, take a moment to pause and reflect on your emotions. Ask yourself why you are feeling a certain way and consider how your emotions may be influencing your judgment.

2. Seek Perspective: Reach out to others for their opinions and insights. Talking through your emotions with someone you trust can provide a fresh perspective and help you make a more balanced decision.

3. Practice Self-Regulation: Develop techniques to manage your emotions effectively. Deep breathing exercises, mindfulness, and journaling are all practical ways to regulate your emotions and maintain clarity during decision-making.

4. Consider Multiple Dimensions: When making decisions, consider not only the rational aspects but also the emotional and ethical dimensions. How do your choices align with your values and long-term goals?

5. Learn from Mistakes: Embrace failure as an opportunity for growth. When emotional decisions lead to less than desirable outcomes, reflect on the experience and learn from it. Use these lessons to inform future decision-making processes.

Real-World Example: Purchasing a Car

Let's illustrate the influence of emotions on decision-making with a real-world example: purchasing a car.

Imagine you have been saving up for your dream car, and you finally have enough money to make the purchase. You visit a dealership and find two options: a practical and fuel-efficient hybrid car or a sleek and sporty luxury car.

While the practical choice may make more sense financially and from a practical standpoint, the luxury car evokes feelings of excitement and status. Your emotions may lead you to choose the luxury car, despite it being less practical.

Understanding the role of emotions in decision-making can help you recognize the influence they have on your choices. By analyzing the situation objectively and considering factors beyond emotions, such as long-term costs and practicality, you can make a decision that aligns with your rational goals and values.

Summary

In this section, we explored the role of emotions in decision-making. We discussed how emotions can influence our judgment and act as decision heuristics. We also highlighted the importance of emotional intelligence in making better decisions and provided strategies for harnessing emotions effectively.

Understanding the interplay between our emotions and rational thinking is crucial for making informed decisions. By balancing our emotions with rational thought, we can navigate decision-making more effectively and achieve outcomes that align with our goals and values.

So, the next time you find yourself faced with an important decision, remember to pause, reflect, and harness the power of your emotions to play smart, not hard!

Exercises

1. Think of a recent decision you made that was influenced by your emotions. Reflect on how your emotions guided your decision-making process. Did they lead to a positive or negative outcome? What could you have done differently to make a more balanced decision?

2. Research and find a real-world example where emotions played a significant role in decision-making. Describe the situation and explain how emotions influenced the outcome. Discuss whether you believe the decision was rational or if emotions played a more dominant role.

3. Consider a situation where you must make a decision that requires both rational thinking and emotional intelligence. How would you approach this decision? What strategies would you use to balance your emotions with rationality?

Analyzing Risk and Uncertainty

In the world of game theory, the concepts of risk and uncertainty play a significant role in decision-making. Understanding how to analyze and navigate these factors is crucial for developing effective strategies. In this section, we will explore the principles of risk and uncertainty and their application in game theory.

Risk vs. Uncertainty

Risk and uncertainty are often used interchangeably, but they have distinct meanings in game theory. Risk refers to situations where the outcome of a decision is unknown, but its probability can be quantified. On the other hand, uncertainty refers to situations where the outcome is unknown, and its probability cannot be precisely determined.

In game theory, risk can be analyzed using probability theory and statistical methods. We can assign probabilities to different outcomes and calculate their expected values. Uncertainty, however, requires a different approach. As it involves unknown or unpredictable factors, we must rely on assumptions, heuristics, and intuition to make decisions.

Expected Utility Theory

One approach to analyzing risk and uncertainty in decision-making is through the lens of expected utility theory. This theory suggests that individuals make choices based on the expected value or utility of each possible outcome.

The expected utility of an outcome is calculated by multiplying the utility of the outcome by its probability and summing up these values for all possible outcomes. Utility represents the individual's subjective satisfaction or preference for a particular outcome.

For example, let's consider a game where you have two options: Option A has a 70% chance of winning 100, and Option B has a 30% chance of winning 200. To calculate the expected utility, we assign utility values to the monetary outcomes and multiply them by their respective probabilities:

$$\text{Expected Utility} = (0.7 \times U(\$100)) + (0.3 \times U(\$200))$$

The decision-maker can then choose the option with the highest expected utility or evaluate the utility values for personal risk tolerance.

Prospect Theory

While expected utility theory provides a framework for decision-making under risk and uncertainty, it has limitations in accurately describing human behavior. Prospect theory, developed by Daniel Kahneman and Amos Tversky, addresses these limitations by incorporating the principles of behavioral economics.

According to prospect theory, individuals do not always make decisions based on expected utility. Instead, they evaluate potential outcomes relative to a reference point and weigh the potential gains and losses differently. The theory suggests that people are risk-averse when facing gains but risk-seeking when facing losses.

For example, if you were given a choice between a guaranteed gain of 100 or a 50% chance of winning 200, most people would choose the guaranteed gain. However, if you were given a choice between a guaranteed loss of 100 or a 50% chance of losing 200, many people would take the gamble to avoid the certain loss.

Prospect theory introduces the concepts of value function and loss aversion. The value function describes how individuals perceive gains and losses, indicating that the perceived value of gains diminishes as they increase, while the perceived value of losses escalates as they deepen.

Decision-Making in the Face of Uncertainty

In situations of uncertainty, decision-making becomes more challenging. While we cannot accurately quantify the probabilities of different outcomes, there are methods and strategies that can help us navigate uncertain scenarios.

One approach is sensitivity analysis, which involves assessing how changes in assumptions or variables impact the decision outcome. By examining a range of possible scenarios and their potential outcomes, decision-makers can gain insights into the level of uncertainty and make more informed choices.

Another strategy is scenario planning, where decision-makers develop multiple plausible future scenarios and evaluate their potential outcomes. This technique allows for flexibility and adaptation, considering various possibilities and developing strategies that can be applied across different scenarios.

Moreover, decision-makers can employ risk mitigation strategies such as diversification, hedging, and insurance to minimize the potential negative impacts of uncertain events. These strategies help spread the risk and provide a safety net in case of unfavorable outcomes.

Real-Life Applications

The principles of risk and uncertainty analysis in game theory find applications in various real-life scenarios. In finance, investment decisions involve evaluating risk and uncertain market conditions. Portfolio managers analyze risk and diversify assets to maximize returns while minimizing potential losses.

In the field of insurance, actuaries use risk assessment models to calculate premiums and estimates based on probabilities and historical data. They evaluate the uncertainty associated with potential claims, ensuring that insurance companies remain financially stable.

Risk analysis is also relevant in the field of public policy. Governments utilize risk management strategies to make decisions regarding public safety, disaster management, and resource allocation.

By analyzing risk and uncertainty, policymakers can develop informed strategies to address complex societal challenges.

Summary

In this section, we explored the concepts of risk and uncertainty in game theory. We learned that risk involves unknown outcomes with quantifiable probabilities, while uncertainty refers to situations where outcomes are unknown and their probabilities cannot be precisely determined. We discussed expected utility theory as a way to analyze risk, considering the utility and probabilities of different outcomes. We also introduced prospect theory, which accounts for the biases and preferences individuals exhibit in decision-making. Finally, we discussed strategies for decision-making in uncertain situations, such as sensitivity analysis, scenario planning, and risk mitigation. The understanding and analysis of risk and uncertainty in game theory provide valuable insights for decision-makers across various fields.

Exercises

1. Think of a real-life decision you recently made that involved both risk and uncertainty. Describe the decision-making process you followed, highlighting how you navigated the elements of risk and uncertainty.

2. Consider the following investment options: Option X has a 60% chance of gaining 10% return and a 40% chance of losing 5% return. Option Y has a 30% chance of gaining 15% return and a 70% chance of losing 8% return. Calculate the expected utility for each option, assuming a linear utility function, and determine which option you would choose.

3. Research and discuss a real-world example where prospect theory explains a decision or behavior that differs from the predictions of expected utility theory. Explain how the principles of prospect theory align with the observed decision or behavior.

4. Imagine you are a government official tasked with developing a disaster management plan for a coastal city vulnerable to hurricanes. Explain how you would use risk and uncertainty analysis to inform your decision-making process, considering the potential outcomes and their associated probabilities.

5. Discuss the limitations of expected utility theory in capturing human decision-making behavior. Provide examples of decision situations where expected utility theory may not accurately represent individuals' preferences and choices.

Prospect Theory: An Alternative to Rational Choice

In the world of game theory, the assumption of rationality is often held as a fundamental principle. However, humans are not always rational creatures, and our decisions are influenced by a multitude of factors beyond pure logic and reason. This is where prospect theory comes into play, offering an alternative perspective to the traditional rational choice model.

Understanding Prospect Theory

Proposed by Amos Tversky and Daniel Kahneman in 1979, prospect theory challenges the notion of rational decision-making by introducing the concept of bounded rationality. According to prospect theory, individuals make decisions based on the potential value of gains and losses, rather than the final outcomes themselves.

Value Function and Psychological Weight

At the core of prospect theory is the value function, which describes how individuals perceive and evaluate outcomes. Unlike the rational choice model, which assumes linear utility functions, prospect theory suggests that people's valuations exhibit diminishing sensitivity to gains and increasing sensitivity to losses.

This concept is captured by the S-shaped value function, where the marginal utility of gains decreases as the value increases, and the marginal utility of losses increases as the value becomes more negative. This asymmetry reflects our tendency to avoid losses more strongly than we seek gains of equal magnitude.

In addition to the value function, prospect theory introduces the concept of psychological weight. These weights represent the relative importance that individuals assign to different outcomes. For example, losses may be perceived as more psychologically significant than equivalent gains, leading to different decision-making patterns.

Reference Points and Framing

Another key aspect of prospect theory is the role of reference points. A reference point serves as a baseline against which outcomes are evaluated. Individuals tend to focus on changes from the reference point rather than the absolute amounts.

Framing, or the way in which information is presented, can significantly influence decision-making under prospect theory. By shifting the reference point or emphasizing certain aspects of a decision, individuals can be swayed to make different choices.

For example, consider a scenario where you have a choice between receiving $100 or flipping a coin for a chance to win $200. If the decision is framed as a potential gain, most people would likely choose the coin flip option due to the possibility of a larger payoff. However, if the decision is framed as a potential loss, individuals may opt for the sure $100 to avoid the risk.

Loss Aversion and Risk Preferences

Prospect theory highlights the concept of loss aversion, which refers to our tendency to strongly dislike losses and be willing to take risks to avoid them. When faced with a potential loss, individuals often exhibit risk-seeking behavior, taking gambles in the hopes of minimizing or reversing the loss.

On the other hand, when it comes to gains, we tend to be risk-averse. The potential psychological pain associated with losing a gain often outweighs the pleasure of making an equivalent gain, leading to a preference for sure gains over risky alternatives.

Bias in Decision-Making

Prospect theory also recognizes various biases that can influence decision-making. One such bias is the framing effect mentioned earlier, where decision outcomes are influenced by how information is presented. Another bias is known as the endowment effect, where individuals tend to value something they own more than they would if they didn't possess it.

Additionally, individuals may exhibit the sunk cost fallacy, continuing to invest time, money, or effort into a losing proposition simply because they have already invested resources into it. These biases can lead to suboptimal decisions and contribute to deviations from rational choice models.

Real-World Applications

Prospect theory has found numerous applications in various fields, ranging from economics and finance to psychology and marketing. For instance, understanding prospect theory can help economists explain why people often hold onto losing stocks in the hope of recouping their losses, despite the availability of potentially more profitable alternatives.

In marketing, prospect theory can help businesses identify effective pricing and promotion strategies by considering consumers' perceptual biases and sensitivity to gains and losses. By framing a product or service as a gain or a loss, marketers can influence consumers' decision-making and increase the likelihood of a purchase.

Exercises

1. Imagine you are the owner of a small business considering a potential investment opportunity. How might prospect theory influence your decision-making process? Consider the potential gains and losses, reference points, and framing effects.

2. Think of a real-life scenario where individuals' decisions were influenced by their aversion to losses. Describe the situation and discuss how loss aversion impacted the choices made.

3. Research and discuss a case study where prospect theory was used to explain or predict consumer behavior. What were the key findings and implications?

Additional Resources

- Tversky, A., & Kahneman, D. (1979). "Prospect Theory: An Analysis of Decision under Risk." *Econometrica*, 47(2), 263-291.

- Thaler, R. H. (1999). *Mental accounting matters.* Journal of Behavioral Decision Making, 12(3), 183-206.

- Kahneman, D. (2011). *Thinking, Fast and Slow.* Farrar, Straus and Giroux.

- Ariely, D. (2008). *Predictably Irrational: The Hidden Forces That Shape Our Decisions.* HarperCollins.

Remember, decision-making is not always rational, and understanding the complexities of prospect theory can provide valuable insights into human behavior and decision processes. So, don't limit yourself to the traditional rational choice model, delve into prospect theory for a deeper understanding of decision-making dynamics.

Bounded Rationality: Making the Best of Limited Information

In the world of decision-making, it's often assumed that humans are rational beings who can gather and process all relevant information to make optimal choices. But let's face it, we're not perfect. Our minds are limited, our information is incomplete, and we have biases that can cloud our judgment. This is where the concept of bounded rationality comes in.

Understanding Bounded Rationality

Bounded rationality is a concept that acknowledges the limitations of human decision-making. Instead of assuming perfect rationality, bounded rationality recognizes that individuals have limited cognitive abilities, time constraints, and access to incomplete information. In other words, we make decisions that are "good enough" rather than optimal.

Herbert Simon, a Nobel laureate in economics, introduced the concept of bounded rationality. According to Simon, decision-makers use heuristic techniques to simplify complex problems and make choices based on the information available. These heuristics are mental shortcuts that allow us to save time and effort, but they can also lead to biases and errors.

Principles of Bounded Rationality

To better understand how bounded rationality works, let's explore some key principles:

1. Satisficing: When faced with a decision, individuals often aim for a satisfactory outcome rather than an optimal one. Instead of trying to find the best solution, they settle for one that meets their minimum criteria. For example, when choosing a hotel room, you might go for one that is reasonably priced and has good reviews, rather than spending hours searching for the absolute best option.

2. **Limited Information Processing:** Bounded rationality recognizes that our minds can only process a limited amount of information at a time. We make decisions based on the information that is readily available to us, without necessarily conducting an exhaustive search or considering all possible alternatives. This can lead to biases and overlook important information.

3. **Cognitive Biases:** Bounded rationality acknowledges that our decision-making is influenced by cognitive biases, which are systematic errors in our thinking. These biases can affect our judgment and lead to irrational choices. Examples of common cognitive biases include confirmation bias (favoring information that confirms our existing beliefs) and availability bias (overestimating the importance of information that is easily accessible).

4. **Sensitive to Context:** Bounded rationality recognizes that decision-making is highly influenced by the context in which choices are made. The same decision might be made differently depending on factors such as time pressure, emotional state, or social norms. We adapt our decision-making strategies based on the specific circumstances we find ourselves in.

The Role of Bounded Rationality in Decision-Making

Bounded rationality is not a limitation; it's a reality. By understanding our cognitive limitations and biases, we can make better decisions despite imperfect information. Here are some strategies to make the most of bounded rationality:

1. **Simplify the Problem:** Break down complex decisions into smaller, manageable parts. Focus on the most relevant information and prioritize what matters most.

2. **Seek Additional Information:** While we have limited information, making an effort to gather more data can improve decision-making. Look for trustworthy sources, consult experts, or seek advice from others who may have a different perspective.

3. **Use Decision-Making Tools:** Various decision-making tools, such as decision matrices or pros and cons lists, can help structure the decision-making process and weigh different factors.

4. **Reflect on Biases:** Be aware of your own cognitive biases and actively challenge them. Consider alternative viewpoints and seek feedback to counterbalance your own subjective biases.

5. Learn from Experience: Reflect on past decisions and outcomes to gain insights. By learning from previous experiences, we can refine our decision-making process and make better choices in the future.

Real-World Examples

Let's explore a couple of real-world examples to illustrate the concept of bounded rationality:

Example 1: Home Buying Decision Imagine you're searching for a new home. You have a limited budget and a long list of criteria. Instead of visiting every available property, you might set a threshold for each criterion. For example, you might consider only homes within a certain price range and with a minimum number of bedrooms. This helps narrow down your choices based on limited information.

Example 2: Online Purchase When shopping online, you encounter numerous options with various characteristics and prices. Rather than meticulously examining every product, you might rely on reviews, ratings, and recommendations to make a decision. These shortcuts allow you to quickly choose a product that meets your basic needs without sifting through all available information.

Remember, bounded rationality doesn't mean we're doomed to make bad decisions. It simply acknowledges the reality of our limited cognitive abilities and imperfect information. By embracing bounded rationality and implementing strategies to mitigate its limitations, we can make informed decisions that are good enough for our needs. So, next time you find yourself faced with a choice, play smart and make the best of your bounded rationality.

Exercises

1. Think of a recent decision you made where you had limited information. Reflect on the outcome and consider if your decision would have been different with more information. How does this align with the concept of bounded rationality?

2. Identify a cognitive bias that you often fall victim to. How does this bias impact your decision-making process? Can you think of strategies to counteract this bias?

3. Choose a decision-making tool (e.g., pros and cons list, decision matrix) and apply it to a recent decision you made. Evaluate how the

tool helped structure your decision-making process and whether it led to a better outcome.

4. Consider a familiar social setting, such as a classroom or workplace. How do the specific context and social dynamics influence decision-making in that setting? Are individuals more likely to rely on bounded rationality due to time constraints or social pressure?

Remember, becoming aware of your own bounded rationality and actively seeking ways to improve your decision-making process will lead to better outcomes in both your personal and professional life.

Further Reading

- Ariely, D. (2008). *Predictably Irrational: The Hidden Forces That Shape Our Decisions*. HarperCollins Publishers. - Kahneman, D. (2011). *Thinking, Fast and Slow*. Farrar, Straus and Giroux. - Simon, H. A. (1997). *Models of Bounded Rationality: Empirically Grounded Economic Reason*. MIT Press. - Thaler, R. H., & Sunstein, C. R. (2008). *Nudge: Improving Decisions About Health, Wealth, and Happiness*. Yale University Press.

These resources provide further insights into the principles of bounded rationality, cognitive biases, and decision-making strategies. Happy reading!

Decision-Making in Competitive and Cooperative Games

In game theory, decision-making plays a crucial role in determining the outcomes of competitive and cooperative games. Understanding how decisions are made can help players strategize and navigate complex game situations. In this section, we will explore the various factors that influence decision-making in both competitive and cooperative game settings.

Rationality and Decision-Making

A fundamental principle in game theory is the assumption of rationality, which postulates that players are utility-maximizers and make choices based on their preferences. Rational decision-making involves carefully weighing the costs and benefits of different actions to select the one that yields the highest expected utility.

However, decision-making under uncertainty can be challenging, and cognitive biases often come into play. These biases can skew judgment and lead to suboptimal decisions. For example, the availability bias causes individuals to rely on recent or easily accessible information, disregarding other relevant factors. The confirmation bias leads people to seek out information that confirms their preconceived beliefs, ignoring contradictory evidence.

The Role of Emotions in Decision-Making

Contrary to the traditional rationality assumption, research has demonstrated that emotions play a crucial role in decision-making. Emotions such as fear, anger, and happiness can influence the evaluation of different options and lead to biased choices.

For instance, in competitive games, anger can fuel aggressive behavior and the desire for retaliation against opponents. Fear, on the other hand, can lead to risk aversion and conservative strategies. Recognizing and managing these emotions can enhance decision-making in competitive game scenarios.

In cooperative games, positive emotions like trust and empathy can foster cooperation and collaboration among players. Decision-making becomes a collective effort aimed at maximizing mutual benefits rather than individual gains. Building strong relationships and maintaining open communication are key to successful cooperative decision-making.

Analyzing Risk and Uncertainty

Risk and uncertainty are inherent in game theory and have a significant impact on decision-making. Risk refers to the known probabilities associated with different outcomes, while uncertainty arises when the probabilities are unknown.

To make decisions in the face of risk and uncertainty, players often rely on expected utility theory. This theory quantifies the desirability of different outcomes by multiplying their respective utilities (or values) by their probabilities and summing them up. By comparing the expected utilities of different actions, players can select the one that maximizes their expected payoff.

However, expected utility theory has been criticized for neglecting the psychological aspects of decision-making and not accounting for individual preferences and attitudes towards risk. Prospect theory, an alternative framework, addresses these limitations by incorporating concepts such as loss aversion and the framing effect into decision-making.

Real-Life Applications of Decision-Making in Game Theory

The principles of decision-making in game theory find numerous real-life applications across various domains. Let's explore a couple of examples:

1. Business Strategies: Companies often face competitive situations where decision-making is critical. For instance, when pricing products, managers must consider competitors' actions and customer demand to maximize profits. By analyzing the strategies adopted by competitors using game theory, companies can make informed decisions and gain a competitive edge.

2. International Relations: Decision-making is also crucial in the realm of international relations, where nations often engage in negotiations and conflicts. By employing game-theoretic models, policymakers can analyze the potential outcomes, risks, and rewards of different courses of action. This understanding guides their decision-making and helps navigate complex diplomatic scenarios.

Exercises

1. Consider a competitive game where two companies are deciding whether to advertise their new products. Each company can choose either to advertise aggressively or conservatively. The payoffs (in millions of dollars) for each possible combination of actions are given by:

Company 1 / Company 2	Aggressive	Conservative
Aggressive	20, 15	10, 25
Conservative	25, 10	15, 20

a) Determine the Nash equilibrium of this game. b) Discuss the decision-making process for each company and the potential psychological factors that may influence their choices.

2. Consider a cooperative game where a group of friends is deciding how to split the cost of a vacation. The total cost of the vacation is $500, and each friend has a different budget constraint. The friends need to

agree on a fair division of the cost. Discuss different decision-making strategies and negotiations that would lead to a stable and fair outcome.

Conclusion

To make effective decisions in competitive and cooperative games, players must consider rationality, emotions, risk, and uncertainty. Understanding these factors and their impact on decision-making allows players to develop strategies that maximize their outcomes. Through real-life applications and exercises, we have seen how decision-making in game theory permeates various fields, from business to international relations. Mastering decision-making in game theory equips individuals with valuable skills for navigating complex strategic interactions.

Group Decision-Making: The Prisoner's Dilemma

In the world of Game Theory, the Prisoner's Dilemma is a classic example that showcases the complexities of decision-making in a group setting. In this section, we will dive into the intricacies of this dilemma, explore its implications, and discuss its real-world examples.

Understanding the Prisoner's Dilemma

Imagine two suspects, Alice and Bob, who have been arrested for a crime but are held in separate cells with no means of communication. The district attorney lacks sufficient evidence to convict them on the main charge, so he offers each of them a plea deal.

The deal goes as follows: if both Alice and Bob confess and testify against each other, they will both be sentenced to a reduced charge of 5 years in prison. If one confesses and the other remains silent, the one who confesses will be set free, while the silent one will face a 10-year sentence. If both remain silent, there is only enough evidence to imprison them for 1 year on a lesser charge.

The prisoners must individually decide whether to cooperate (remain silent) or defect (confess). The dilemma arises when they attempt to maximize their own self-interest while considering the potential outcomes for both themselves and the other person.

Solving the Prisoner's Dilemma

To analyze the Prisoner's Dilemma mathematically, we assign the following payoffs:

- If both prisoners cooperate (remain silent), they each receive a payoff of 1.

- If both prisoners defect (confess), they each receive a payoff of -5.

- If one prisoner defects while the other cooperates, the defector receives a payoff of 0, while the cooperator receives a payoff of -10.

We can represent these payoffs in a payoff matrix:

	Cooperate (Remain Silent)	Defect (Confess)
Cooperate (Remain Silent)	1, 1	-10, 0
Defect (Confess)	0, -10	-5, -5

To find the Nash Equilibrium, which represents the most stable outcome, we look for the best responses for each player. In this case, both players defect (confess) since it yields a higher payoff than cooperating (remaining silent), regardless of the other player's choice.

Thus, the Nash Equilibrium in the Prisoner's Dilemma is for both players to defect, resulting in a suboptimal outcome for both individuals. This shows that self-interest can lead to a worse collective outcome.

Real-World Applications

The Prisoner's Dilemma can be observed in various real-world scenarios, shedding light on the challenges of group decision-making. Let's explore a few examples:

1. **Arms Race:** Countries engaging in an arms race must decide whether to develop more weapons (defect) or cooperate and focus on peacebuilding efforts (cooperate). The fear of being left defenseless often leads to a defection strategy, resulting in an arms race that benefits no one.

2. **Business Competition:** In a competitive industry, companies face a dilemma when deciding whether to collaborate (cooperate) or implement aggressive tactics like price wars (defect). While cooperation may lead to long-term benefits, the temptation to gain a competitive advantage often prevails, resulting in a detrimental cycle of aggressive actions.

3. **Climate Change:** When it comes to mitigating climate change, countries must choose between reducing their carbon emissions (cooperate) or exploiting resources without regarding environmental consequences (defect). The dilemma lies in the fact that if one country defects and takes advantage of resources, while others cooperate, they may gain a short-term advantage but contribute to global harm in the long run.

These examples demonstrate the relevance of the Prisoner's Dilemma in various fields and highlight the challenges of achieving collective cooperation in the face of self-interest.

Ethics and Morality

The Prisoner's Dilemma raises profound ethical questions about the nature of decision-making and moral responsibility. It challenges individuals to consider the consequences of their actions not only for themselves but also for the collective.

In the Prisoner's Dilemma, the defecting strategy may appear rational from a self-interest perspective, but it neglects the potential for mutual benefit. By prioritizing short-term gains, individuals risk perpetuating a cycle of suboptimal outcomes.

It is important to recognize that cooperation is essential for creating a better society. By embracing cooperative strategies, individuals can break free from the limitations imposed by the Prisoner's Dilemma and foster a more harmonious and prosperous collective.

Exercise: The International Climate Agreement

Consider the following scenario: A group of countries must decide whether to join an international climate agreement. If all countries cooperate and reduce their carbon emissions, the global community will benefit through reduced climate change impacts. However, if a few countries defect and exploit resources without regard for the environment, they may gain a temporary economic advantage.

Using the concepts of the Prisoner's Dilemma, analyze the potential outcomes and discuss the challenges of reaching a cooperative solution in this scenario. Consider the factors that might influence countries' decisions and propose strategies for fostering cooperation.

Resources

To further explore the complexities of group decision-making and the implications of the Prisoner's Dilemma, consider the following resources:

- Book: "The Evolution of Cooperation" by Robert Axelrod
- Article: "The Tragedy of the Commons" by Garrett Hardin
- Video: "The Prisoner's Dilemma Explained" by TED-Ed
- Online Game: "The Evolution of Trust" by Nicky Case

These resources offer valuable insights into the dynamics of decision-making and encourage critical thinking about cooperation and self-interest.

Conclusion

The Prisoner's Dilemma serves as a powerful framework for understanding the challenges of group decision-making. It highlights the tension between self-interest and collective well-being, emphasizing the need for cooperation in creating optimal outcomes.

By recognizing the limitations of a purely individualistic approach, we can strive to overcome the Prisoner's Dilemma and envision a future where cooperation prevails over defection. As we continue to navigate complex societal issues, understanding game theory and its implications remains crucial for making smarter decisions, both individually and as a group.

Behavioral Economics and Game Theory

Behavioral economics is a field that integrates insights from psychology into the study of economic decision-making. It challenges the traditional assumption of "rationality" in economics and explores how individuals actually make choices in real-world situations. In the context of game theory, behavioral economics provides a more realistic understanding of human behavior and its impact on strategic interactions.

Bounded Rationality: Beyond Perfect Rationality

Traditional game theory assumes that individuals are perfectly rational, meaning that they make consistent decisions based on complete and accurate information. However, behavioral economists challenge this assumption by introducing the concept of bounded rationality. Bounded rationality recognizes that individuals have limited cognitive abilities and must make decisions under conditions of uncertainty.

In the context of game theory, bounded rationality acknowledges that players may not always make optimal choices due to cognitive limitations. They may rely on rules of thumb, heuristic strategies, or even be influenced by emotions, biases, and social factors. These deviations from perfect rationality have significant implications for strategic interactions and can lead to outcomes that differ from the predictions of traditional game theory.

Cognitive Biases and Decision-Making

Cognitive biases are systematic patterns of thinking that often lead to deviations from rational decision-making. These biases play a critical role in game theory as they influence how individuals perceive and evaluate strategic situations. Let's explore a few common cognitive biases and their impact on decision-making in games:

- **Confirmation Bias:** The tendency to interpret information in a way that confirms preexisting beliefs or expectations. In games, this bias can lead players to selectively focus on evidence that supports their preferred strategies, ignoring information that challenges their beliefs.

- **Loss Aversion:** The tendency to dislike losses more than equivalent gains. Loss aversion can make players more risk-averse, leading them to choose suboptimal strategies in order to avoid potential losses.

- **Overconfidence Bias**: The tendency to overestimate one's own abilities or the accuracy of one's own beliefs. Overconfident players may make overly optimistic predictions about the outcomes of games, leading them to adopt aggressive strategies that may not be optimal.

- **Anchoring Bias**: The tendency to rely heavily on the first piece of information encountered when making decisions. In games, anchoring bias can influence players' initial strategy choices, potentially leading to suboptimal outcomes.

By understanding and accounting for these biases, game theorists can develop more accurate models of decision-making and provide better predictions for how individuals will behave in strategic situations.

Experimental Game Theory

To study the impact of behavioral factors on strategic interactions, game theorists often employ experimental methods. Experimental game theory involves creating controlled laboratory settings where participants play games and their decisions are recorded and analyzed.

Experimental game theory allows researchers to test various hypotheses about human behavior and explore the role of cognitive biases, social preferences, and cultural factors in decision-making. By observing participants' choices, researchers can derive empirical data and compare it to the predictions of traditional game theory models.

For example, the famous Ultimatum Game is often used in experiments to study fairness and bargaining behavior. In this game, one player proposes a division of a sum of money, and the other player can either accept or reject the offer. If the offer is rejected, both players receive nothing. Traditional game theory predicts that players will act rationally and accept any positive offer, even if it is unequal. However, experimental evidence consistently shows that players tend to reject unfair offers, suggesting that notions of fairness and reciprocity are important drivers of behavior.

Nudging and Behavioral Interventions

Behavioral economics also explores ways to influence decision-making and shape behavior through the use of "nudging" and behavioral interventions. Nudging refers to designing the choice architecture in a way that encourages individuals to make certain decisions without restricting their freedom of choice.

In the context of game theory, nudging can be used to align individual incentives with desirable outcomes. For example, by promoting social norms that encourage cooperation, game theorists can incentivize players to choose cooperative strategies rather than defecting. Nudging can also be used to mitigate the impact of cognitive biases by providing individuals with helpful cues or reminders.

Additionally, behavioral interventions can be deployed to address market failures or encourage socially beneficial behaviors. For example, by providing information about the energy consumption of household appliances, individuals can make more informed decisions and reduce energy consumption.

Real-World Applications of Behavioral Economics and Game Theory

Behavioral economics and game theory have numerous real-world applications across various domains. Here are a few examples:

- **Public policy**: Behavioral economics informs the design of policies aimed at improving public welfare. By understanding how individuals make decisions, policymakers can develop interventions that encourage desirable behaviors, such as saving for retirement or reducing harmful habits like smoking.

- **Marketing and advertising**: Behavioral economics provides insights into consumer behavior, allowing marketers to design effective advertising strategies. By understanding cognitive biases and decision-making processes, marketers can create persuasive messages that influence consumer choices.

- **Financial markets**: Understanding behavioral biases is crucial for predicting market reactions and designing investment strategies. By incorporating behavioral factors into financial models, analysts can better understand market dynamics and identify investment opportunities.

- **Healthcare:** Behavioral economics plays a significant role in public health interventions. By understanding the factors that influence health-related behaviors, policymakers can design interventions to promote healthy habits, such as exercising regularly or maintaining proper nutrition.

Conclusion

Behavioral economics enriches game theory by providing a more nuanced understanding of human behavior in strategic interactions. By acknowledging bounded rationality, cognitive biases, and the role of social factors, game theorists can develop more accurate models and predictions. Experimental methods and behavioral interventions further enrich the field, allowing researchers to observe and influence decision-making in controlled settings. The real-world applications of behavioral economics and game theory span various domains, from public policy to marketing and finance, making it an essential field for understanding and improving human interactions. So, fuck traditional assumptions of rationality and embrace the complex reality of human decision-making!

Exercise: Think about a recent strategic decision you made. Can you identify any cognitive biases or behavioral factors that influenced your choice? How might understanding these biases have influenced your decision differently?

Real-Life Applications of Decision-Making in Game Theory

In the previous section, we explored various aspects of decision-making in game theory, from rational choice models to cognitive biases and emotions. Now, let's delve into the real-life applications of these concepts and how they shape the decisions we make in different fields.

Economics

Game theory has had a profound impact on the field of economics, providing valuable insights into decision-making and strategic interactions. One prominent application is in understanding oligopolies, where a small number of firms dominate the market. Through game theory, economists can analyze the behaviors of these firms and predict outcomes in terms of pricing and competition.

For example, consider the airline industry, which is characterized by a few major airline companies. To attract customers, each airline must decide on ticket prices, taking into account the potential reactions of their competitors. Game theory helps us understand how these decisions are made and the resulting market equilibrium.

Game theory is also used in auctions, where bidders strategically determine their bids to maximize their chances of winning while minimizing their costs. Different auction formats, such as first-price and second-price auctions, have distinct strategies associated with them. Understanding these strategies allows auctioneers to design efficient auctions that generate optimal outcomes.

Politics

Game theory has been applied extensively in political science to analyze strategic decision-making and policy choices. In international relations, countries often engage in negotiations and conflicts, which can be modeled using game theory.

One well-known example is the Prisoner's Dilemma, where two suspects are interrogated and face the choice of cooperating or betraying each other. This dilemma highlights the tension between individual incentives and collective outcomes. By utilizing game theory, researchers can better understand why cooperation sometimes fails and how trust can be established in international relations.

Voting theory is another domain where game theory reveals its relevance. Different voting systems, such as plurality voting and ranked-choice voting, have distinct strategic implications. Game theory helps us analyze outcomes, such as strategic voting and coalition formation, which shape the democratic process.

Negotiations

In everyday life, negotiations are an integral part of decision-making, whether it's regarding salary negotiations, business deals, or even personal relationships. Game theory provides valuable insights into the dynamics of negotiations and helps identify optimal strategies.

One classic example is the Ultimatum Game, where two players must agree on how to divide a sum of money. The proposer suggests a division, and the responder must decide whether to accept or reject it. Game theory allows us to understand why proposers offer certain splits and how responders perceive fairness. It also helps predict the outcomes based on the participants' strategic calculations.

In business negotiations, game theory can be applied to analyze bargaining power and the possibilities of reaching mutually beneficial agreements. This understanding of strategic behavior enables negotiators to optimize their outcomes and achieve better results.

Environmental Science

Game theory has also found applications in environmental science, particularly in understanding international efforts to address collective environmental issues. For example, the management of shared resources, such as fisheries or water basins, involves multiple stakeholders who must make decisions that balance their own interests with the common good.

By utilizing game theory, researchers can analyze the interactions between these stakeholders and identify strategies that promote cooperation and sustainable resource management. This can lead to more effective policies for conserving natural resources and mitigating environmental challenges.

Social Behavior

The study of social behavior is another area where game theory plays a significant role. By using game theory, researchers can model and understand various social situations, such as cooperation, trust, and reciprocity.

For instance, the Prisoner's Dilemma can be applied in social behavior experiments to explore the incentives for cooperation or defection. It helps shed light on the factors that promote or hinder prosocial behavior in society.

Game theory is also used to examine social networks, where individuals are connected through various relationships. These networks exhibit complex dynamics, and game theory provides tools to analyze how behaviors spread and influence each other within the network. Such analysis can reveal important insights into phenomena like the spread of ideas or opinions through social media platforms.

Summary

In this section, we explored the real-life applications of decision-making in game theory across various fields. From economics to politics, negotiations to environmental science, and social behavior to social networks, game theory provides a powerful framework for understanding and predicting strategic interactions.

By applying game theory principles in practical situations, we can make informed decisions, anticipate the actions of others, and navigate complex scenarios more effectively. As the world becomes increasingly interconnected and decisions become more strategic, understanding game theory and its applications becomes ever more crucial.

Remember, smart decision-making isn't just about working hard—it's about playing smart. And game theory equips us with the knowledge and tools to do just that. So let's continue our journey and explore the intriguing world of types of games and their solutions in the next section.

Types of Games

Two-Player Zero-Sum Games

The Basics of Two-Player Zero-Sum Games

Game theory is all about understanding strategic decision-making in competitive situations, and two-player zero-sum games are a fundamental concept that lies at the heart of this field. In this section, we will explore the basics of two-player zero-sum games, including their definition, strategies, and equilibrium solutions.

Definition

A two-player zero-sum game is a type of game where the interests of the players are directly opposed to each other. In other words, whatever one player gains, the other player loses, and the total payoff of the game is constant. It's like a seesaw where one player's gain is the other player's loss, and vice versa.

To formalize the concept, let's consider a game matrix, also known as a payoff matrix. In this matrix, the rows represent the strategy choices of Player 1, and the columns represent the strategy choices of Player 2. The entries in the matrix represent the payoffs, which can be positive (a gain for Player 1) or negative (a loss for Player 1). Since it is a zero-sum game, the sum of payoffs across all cells is always zero.

Let's take a classic example: rock-paper-scissors. In this game, Player 1 chooses between rock, paper, or scissors, and Player 2 also chooses between the same options. The payoff matrix for this game would look like this:

	Rock	Paper	Scissors
Rock	0	-1	1
Paper	1	0	-1
Scissors	-1	1	0

In this matrix, the entry (-1,1) represents Player 1 choosing rock and Player 2 choosing paper, resulting in a gain of 1 for Player 2 and a loss of 1 for Player 1. Notice how the sum of payoffs in each cell always adds up to zero.

Strategies

In two-player zero-sum games, players strategize to maximize their own payoff while minimizing their opponent's payoff. Each player must choose a strategy, which is a complete plan of action that specifies their move for any possible move by the opponent.

In our rock-paper-scissors example, a strategy could be simply choosing each option with equal probabilities. This strategy is called a mixed strategy because it involves randomness. Another type of strategy is a pure strategy, where a player chooses a specific option with certainty. For example, choosing rock every time would be a pure strategy.

Nash Equilibrium

The concept of Nash equilibrium is central to the analysis of two-player zero-sum games. A Nash equilibrium is a set of strategies where no player can unilaterally deviate and improve their own payoff. In other words, it's a stable state where both players are playing optimally given the other player's strategy.

To find the Nash equilibrium in a two-player zero-sum game, we need to identify any strategies where the players have no incentive to change their moves. This occurs when there is a pair of strategies, one for each player, such that the payoffs for each player are optimized against the other player's strategy.

In the rock-paper-scissors example, there is no pure strategy Nash equilibrium because each option can be beaten by another. However, there is a mixed strategy Nash equilibrium, where each player chooses their options with equal probabilities. In this equilibrium, both players have no incentive to change their strategy because any deviation would result in a lower expected payoff.

Applications

Two-player zero-sum games have numerous applications in economics, politics, and other fields. One example is in the study of auctions, where bidders compete to buy or sell goods. Auction theory uses game theory to analyze different auction formats, optimal bidding strategies, and the determination of auction prices.

Another application is in international relations, where countries engage in strategic interactions such as negotiations and conflicts. Game theory helps analyze the decision-making processes in these situations, predicting outcomes and understanding how cooperation and conflict arise.

Summary

In this section, we explored the basics of two-player zero-sum games. We learned that these games involve direct competition between two players, where one player's gain is the other player's loss. We discussed strategies, including mixed and pure strategies, and introduced the concept of Nash equilibrium as the stable state of optimal play. Finally, we highlighted some applications of two-player zero-sum games in real-world contexts like auctions and international relations.

Now that we have a solid understanding of the fundamentals, let's dive deeper into different types of games and their strategies in the following sections. Remember, game theory is not just about winning but also about understanding and reasoning strategically in a competitive world. So, let's play smart, not hard!

Minimax Strategy: Maximizing Gain and Minimizing Loss

In the exciting world of game theory, where outcomes are determined by strategic interactions, players often face the challenge of making decisions that maximize their gains while minimizing their losses. One powerful tool that players can employ to achieve this is the minimax strategy.

The minimax strategy is a fundamental concept in game theory and is particularly useful in two-player zero-sum games. In this type of game, whatever one player gains, the other must lose, hence the term "zero-sum." Examples of such games include chess, poker, and tic-tac-toe.

The main idea behind the minimax strategy is for each player to anticipate the best possible move for themselves while considering the worst possible move their opponent could make. By planning their strategy in this way, players aim to minimize the potential losses they may face, assuming their opponent will also make optimal moves.

To illustrate how the minimax strategy works, let's consider a simple example of a two-player zero-sum game: Rock-Paper-Scissors. In this game, each player simultaneously chooses one of three options: rock, paper, or scissors. The outcomes are determined by the following rules:

- Rock beats scissors
- Scissors beat paper
- Paper beats rock
- If both players choose the same option, it's a tie

To apply the minimax strategy, each player must consider the worst-case outcome for themselves based on the choices their opponent could make. In Rock-Paper-Scissors, the worst-case outcome for a player is a tie, as it doesn't result in a gain but also avoids a loss. By analyzing all possible moves and their potential outcomes, players can make informed decisions to maximize their chances of winning.

Let's say Player A is using the minimax strategy and is about to make their move. They will choose the option that minimizes their potential losses, assuming Player B will also make the best move possible. Similarly, Player B is also employing the minimax strategy and will select the option that minimizes their potential losses, assuming Player A will do the same.

By employing the minimax strategy, each player strives to maximize their gains while minimizing their losses. In a game like Rock-Paper-Scissors, this strategy may not guarantee victory every time since there is an element of chance involved. However, in games with perfect information and deterministic outcomes, the minimax strategy can lead to an optimal outcome.

It's important to note that the minimax strategy assumes rationality from all players, meaning they will make decisions based on their best interests. Additionally, to ensure fairness and an optimal outcome, the players must have complete knowledge of the game's rules and be aware of all possible moves and their outcomes.

The minimax strategy has been widely studied and applies to various fields, including economics, psychology, and computer science. In economics, it helps analyze competitive markets, while in psychology, it sheds light on decision-making processes. Furthermore, in computer science, the minimax strategy forms the foundation for designing artificial intelligence algorithms that can make optimal decisions in games.

In summary, the minimax strategy is a powerful tool that enables players to make decisions that maximize their gains while minimizing their losses in two-player zero-sum games. By considering the worst-case scenario and anticipating their opponents' moves, players can select strategies that provide them with the greatest advantage. Whether you're playing a friendly game of Rock-Paper-Scissors or exploring the complexities of strategic interactions, the minimax strategy is a valuable concept to understand and apply.

Key Takeaways:

- The minimax strategy is an important concept in game theory, particularly in two-player zero-sum games.

- It involves considering the worst-case outcome for oneself while assuming the opponent will make optimal moves.

- The strategy aims to minimize potential losses and maximize gains for each player.

- The minimax strategy is based on the assumption of rationality and complete knowledge of the game.

- It has applications in economics, psychology, and computer science, among other fields.

Exercises:

1. Consider a game of chess. Suppose you are the white player, and it's your turn to move. Use the minimax strategy to determine your

best move, assuming your opponent will make optimal moves as well.

2. Think of a real-world scenario where the minimax strategy could be applied. Describe the situation, the players involved, and how the minimax strategy could help maximize gains and minimize losses.

3. Research and find an example of a real-world competitive market. Analyze the market using the minimax strategy framework. What strategies could the players employ to maximize their gains?

Additional Resources:

- Dixit, A., & Nalebuff, B. (2008). *The Art of Strategy: A Game Theorist's Guide to Success in Business and Life*. W. W. Norton & Company.

- Binmore, K. (2007). *Playing for Real: A Text on Game Theory*. Oxford University Press.

- Osborne, M. J., & Rubinstein, A. (1994). *A Course in Game Theory*. MIT Press.

So next time you find yourself facing a strategic decision, remember the minimax strategy and play smart, not hard!

Nash Equilibrium: Finding Stability in Two-Player Zero-Sum Games

In this section, we dive into the concept of Nash equilibrium, a fundamental concept in game theory that holds the key to finding stability in two-player zero-sum games. We will explore how players can strategically choose their actions to maximize their outcomes and how Nash equilibrium provides a state of balance where no player has an incentive to deviate from their strategy unilaterally.

Understanding Nash Equilibrium

To understand Nash equilibrium, let's start with the basics. In a two-player zero-sum game, the utility or payoff for one player is directly opposite to the other player's. This means that whatever one player gains, the other player loses, and vice versa. Examples of such games include chess, poker, or even rock-paper-scissors.

TWO-PLAYER ZERO-SUM GAMES

Nash equilibrium is a concept coined by John Nash, a Nobel laureate in economics and mathematician. It refers to a state in the game where each player, knowing the strategies of the others, chooses the best strategy for themselves. In other words, no player has an incentive to unilaterally deviate from their strategy, given the strategies of the other players.

Finding Nash Equilibrium

Finding the Nash equilibrium in a game involves determining the best strategy for each player, given the strategies chosen by others. This can be done through a process of elimination or by using mathematical equations.

Let's consider a simple example: the game of matching pennies. In this game, Player 1 and Player 2 each choose to show either heads or tails by placing a penny, heads up or tails up, simultaneously. Player 1 wins if the pennies match (both heads or both tails), and Player 2 wins if the pennies do not match.

To find the Nash equilibrium, we need to analyze the strategies of both players and determine if there is a stable outcome.

Player 1's Strategies: Player 1 has two choices: heads (H) or tails (T). Let's assume that Player 1 chooses heads with a probability p and tails with a probability $1 - p$.

Player 2's Strategies: Similarly, Player 2 also has two choices: heads (H) or tails (T). Let's assume that Player 2 chooses heads with a probability q and tails with a probability $1 - q$.

Now, we need to analyze the payoffs for each player given the strategies chosen by themselves and their opponent.

Payoff Analysis:

	Player 2: H	Player 2: T
Player 1: H	1	-1
Player 1: T	-1	1

If Player 1 chooses heads (H) and Player 2 chooses heads (H), Player 1 wins with a payoff of 1, and Player 2 loses with a payoff of -1. Similarly, if Player 1 chooses tails (T) and Player 2 chooses tails (T), Player 1 loses with a payoff of -1, and Player 2 wins with a payoff of 1.

To find the Nash equilibrium, we need to determine if any changes in strategy would benefit either player. Let's analyze the possible scenarios:

- If Player 1 deviates from their chosen probability p towards heads (H), Player 2 can increase their probability q of choosing heads (H) to maximize their payoff. The same applies if Player 1 deviates towards tails (T).

- If Player 2 deviates from their chosen probability q towards heads (H), Player 1 can adjust their probability p of choosing heads (H) to maximize their payoff. The same applies if Player 2 deviates towards tails (T).

Since both players have optimal strategies in place, there is no incentive for either player to unilaterally deviate. In this case, the Nash equilibrium occurs when both players choose their strategies with equal probabilities: $p = q = 0.5$.

Real-World Applications

Nash equilibrium finds application in various fields, including economics, politics, biology, and social interactions. It provides insights into human behavior, decision-making, and strategic interactions.

One prominent application is in the analysis of oligopolies, where a few firms dominate the market. Each firm's pricing decisions depend on the strategies of their competitors. The Nash equilibrium helps determine the stable price levels where no firm has an incentive to change its price unilaterally.

Another application is in traffic flow optimization, where drivers choose between different routes. By understanding the Nash equilibrium, transportation planners can predict the distribution of traffic and propose policies that minimize congestion.

Furthermore, Nash equilibrium also finds significance in conflict resolution and negotiation strategies. By identifying stable equilibria in negotiations, parties can reach mutually beneficial outcomes and avoid unnecessary conflicts.

Caveats and Unconventional Insights

It is important to note that Nash equilibrium assumes rationality and complete information on the part of players. However, in real-world scenarios, players may have limited information, face cognitive biases, or act irrationally. These factors can significantly affect the stability of Nash equilibria.

Additionally, the existence of multiple Nash equilibria is possible in certain games, making the analysis more complex. In such cases, players may need to reason about their opponents' reasoning, leading to a concept called higher-order beliefs.

One unconventional insight is that Nash equilibrium does not always guarantee the best outcome for all players. It merely represents a state of stability where no player has the incentive to deviate. Hence, in some cases, players may be stuck in suboptimal equilibria, missing out on better outcomes.

Summary

Nash equilibrium is a powerful concept in game theory that provides stability and strategic insights in two-player zero-sum games. By analyzing players' strategies and payoffs, we can determine stable outcomes where no player has an incentive to unilaterally deviate. Nash equilibrium finds application in various real-world scenarios, including economics, politics, and everyday social interactions. However, it is essential to consider the limitations and deviations from rationality that may impact the accuracy of Nash equilibria in real-life situations.

Mixed Strategies: Adding Uncertainty to the Game

In the previous section, we explored the basics of two-player zero-sum games and learned how to find the optimal strategies using concepts like minimax strategy and Nash equilibrium. However, in many real-world scenarios, players don't always have a single pure strategy that guarantees the best outcome. This is where mixed strategies come into play.

Introduction to Mixed Strategies

In game theory, a mixed strategy refers to a player's strategy that involves randomizing their actions based on a certain probability distribution. Unlike pure strategies, which involve a deterministic choice of actions, mixed strategies introduce an element of uncertainty to the game, making it more realistic and reflective of real-life decision-making processes.

To understand the concept of mixed strategies, let's consider a simple example: the game of rock-paper-scissors. In this game, players simultaneously choose one of the three options (rock, paper, or scissors). If both players choose the same option, it's a tie; otherwise, rock beats scissors, scissors beat paper, and paper beats rock.

Now, suppose player A decides to play rock with a probability of p, paper with a probability of q, and scissors with a probability of $1-p-q$, where $0 \leq p, q \leq 1$ and $p+q \leq 1$. Similarly, player B decides to play rock with a probability of r, paper with a probability of s, and scissors with a probability of $1-r-s$. This assignment of probabilities represents the mixed strategies of both players.

Finding Expected Payoffs

To analyze the game with mixed strategies, we need to determine the expected payoffs for each player. The expected payoff is the anticipated value that a player can expect to receive by playing a particular strategy, taking into account the probabilities assigned to the different pure strategies.

In the game of rock-paper-scissors, the payoffs can be represented in a matrix as follows:

	Rock	Paper	Scissors
Rock	$(0,0)$	$(-1,1)$	$(1,-1)$
Paper	$(1,-1)$	$(0,0)$	$(-1,1)$
Scissors	$(-1,1)$	$(1,-1)$	$(0,0)$

Here, the first value in each cell represents the payoff for player A, while the second value represents the payoff for player B.

To find the expected payoffs, we multiply the probabilities of each player with the corresponding payoffs and sum them up. For player A, the expected payoff can be calculated as:

$$E(A) = p \cdot (0) + q \cdot (-1) + (1-p-q) \cdot (1)$$

Similarly, for player B, the expected payoff can be calculated as:

$$E(B) = r \cdot (0) + s \cdot (1) + (1-r-s) \cdot (-1)$$

By solving these equations, we can obtain the expected payoffs for both players, which will give us insights into the best strategies to adopt.

Analysis and Interpretation

Once we have the expected payoffs for each player, we can determine the best strategies to adopt. In mixed strategy equilibrium, both players are indifferent between their available pure strategies, given the mixed strategies being played by the other player.

To find the mixed strategy equilibrium, we equate the expected payoffs for each pure strategy and solve the resulting equations. For example, equating the expected payoffs for player A's choices of rock and paper, we get:

$$p(0) + q(-1) + (1 - p - q)(1) = p(-1) + q(0) + (1 - p - q)(-1)$$

By solving this equation, we can determine the mixed strategies that lead to equilibrium.

In the case of rock-paper-scissors, the mixed strategy equilibrium occurs when all three options are played with equal probabilities: $p = q = r = s = \frac{1}{3}$. This means that both players should randomize their choices equally among rock, paper, and scissors to maximize their expected payoffs.

Real-World Applications

Mixed strategies have significant applications in various real-world scenarios. Some examples include:

- **Business competition:** In industries where firms have multiple products or pricing options, they can adopt mixed strategies to maximize market share or profits.

- **Sports strategies:** Coaches in team sports often use mixed strategies to select plays or formations, making it difficult for opposing teams to predict their moves.

- **Political campaigns:** Politicians may use mixed strategies to decide where to allocate campaign resources or target different demographics effectively.

- **Military tactics:** In warfare, mixed strategies can be employed to create uncertainty and confusion among adversaries, making it harder for them to anticipate the next move.

By incorporating uncertainty into decision-making processes, mixed strategies allow players to adapt and respond to dynamic situations, maximizing their chances of success.

Summary

In this section, we explored the concept of mixed strategies in game theory. Mixed strategies introduce uncertainty and randomness into the decision-making process, reflecting real-world scenarios where players don't always have a single deterministic strategy. We learned how to find the expected payoffs for each player when mixed strategies are involved and how to identify the mixed strategy equilibrium. Finally, we explored real-world applications of mixed strategies in various fields.

Now that we have a solid understanding of mixed strategies, we can move on to the next section, where we will delve into the fascinating world of cooperative games. Stay tuned!

Exercise: Consider a two-player game with the following payoff matrix:

	Pure Strategy 1	Pure Strategy 2
Pure Strategy 1	$(3, 2)$	$(1, 4)$
Pure Strategy 2	$(2, 3)$	$(4, 1)$

(a) Calculate the expected payoffs for each player when they employ mixed strategies.
(b) Determine the mixed strategy equilibrium for this game.
(c) Can you think of a real-world scenario that corresponds to this game?

Resources:

- Dixit, A., and Nalebuff, B. (2008). *The Art of Strategy: A Game Theorist's Guide to Success in Business and Life.* W. W. Norton & Company.

- Binmore, K. (2007). *Game Theory: A Very Short Introduction.* Oxford University Press.

- *Game Theory: Mixed Strategies.* (n.d.). Retrieved from https://www.economicsonline.co.uk/Competitive_markets/Game_theory_mixed_strategies.html

Remember, life is a game, and learning game theory is your secret weapon. Keep hustling, stay sharp, and always play smart, not hard!

Linear Programming and Game Theory

Linear programming is a mathematical technique that helps us make optimal decisions in situations where we have limited resources and need to achieve specific objectives. It is a powerful tool that is widely used in various fields, including economics, operations research, management science, and game theory.

In the context of game theory, linear programming provides a framework for analyzing and solving games with multiple players and conflicting objectives. It allows us to determine the best strategies for each player, considering the constraints imposed by the game's structure and the players' objectives.

To understand the connection between linear programming and game theory, let's dive into some key concepts and techniques.

The Basics of Linear Programming

Linear programming is based on the principles of optimization. It involves manipulating a set of linear equations or inequalities to find the values of variables that optimize a given objective function, subject to certain constraints.

In a linear programming problem, we have a set of decision variables that we want to determine. These variables represent the quantities or actions we can control to achieve our desired objective. We also have an objective function, which is a linear equation that defines what we want to optimize. The constraints are linear inequalities that represent the limitations or restrictions on the variables.

The solution to a linear programming problem is a set of values for the decision variables that satisfy all the constraints and maximize or minimize the objective function.

Applying Linear Programming to Game Theory

In game theory, we can use linear programming to analyze and solve certain types of games, such as two-player zero-sum games and cooperative games.

Let's start by considering two-player zero-sum games. In these games, the total payoff for one player is equal to the total loss for the other player. The players have conflicting objectives, and their strategies are directly opposed to each other.

Linear programming can help us find the optimal strategies for each player in these games. We can set up a linear programming problem where the decision variables represent the probability or frequency of choosing each possible strategy. The objective function captures the payoff or utility for each player, and the constraints ensure that the probabilities sum to 1 and are non-negative.

By solving the linear programming problem, we can find the Nash equilibrium, which is a stable solution where neither player has an incentive to unilaterally deviate from their strategy. The Nash equilibrium represents the optimal strategies for both players in the game.

An Example: The Prisoner's Dilemma

To illustrate the application of linear programming in game theory, let's consider the classic example of the Prisoner's Dilemma.

In the Prisoner's Dilemma, two suspects are held in separate cells and face the decision to either cooperate with each other or betray each other. Depending on their choices, they will receive different prison sentences. The payoff matrix for the Prisoner's Dilemma can be represented as follows:

	Cooperate	Betray
Cooperate	(-1, -1)	(-3, 0)
Betray	(0, -3)	(-2, -2)

In this game, the first number in each cell represents the payoff for the suspect in the first row, and the second number represents the payoff for the suspect in the first column.

To find the optimal strategies for the suspects, we can set up a linear programming problem. Let's define the decision variables as x and y, representing the probabilities of cooperating for the first and second suspects, respectively. The objective function is $-x - y$, which represents the total payoff for the suspects. The constraints ensure that the probabilities sum to 1 and are non-negative: $x + y = 1$ and $x, y \geq 0$.

By solving this linear programming problem, we can find the optimal strategies for the suspects. In this case, the optimal solution is $x = \frac{2}{5}$ and $y = \frac{3}{5}$. This means that the suspects should cooperate with a probability of $\frac{2}{5}$ and betray with a probability of $\frac{3}{5}$.

The Nash equilibrium in the Prisoner's Dilemma is for both suspects to betray each other, resulting in a payoff of $(-2, -2)$.

Real-World Applications

Linear programming and game theory have numerous real-world applications. For example, in business and economics, they can be used to optimize production and distribution decisions, determine pricing strategies, and analyze market competition.

In operations research, linear programming is used to optimize resource allocation, such as workforce scheduling or inventory management.

In political science and international relations, game theory and linear programming can help analyze decision-making in conflicts, negotiations, and alliances.

Linear programming is a versatile tool that can be applied in various domains to make informed decisions and understand strategic interactions.

Further Reading and Resources

If you want to explore linear programming and its applications in more depth, here are some recommended resources:
   - "Introduction to Operations Research" by Frederick S. Hillier and Gerald J. Lieberman - "Game Theory: An Introduction" by Steven Tadelis - "Game Theory: Analysis of Conflict" by Roger B. Myerson - "Linear Programming and Network Flows" by Mokhtar S. Bazaraa, John J. Jarvis, and Hanif D. Sherali

Additionally, there are various online courses and tutorials available on platforms like Coursera, edX, and Khan Academy that can help you develop a deeper understanding of linear programming and game theory.

Remember, linear programming is just one tool in the vast field of game theory. By combining it with other concepts and techniques, we can gain valuable insights into decision-making, strategic interactions, and cooperative behavior.

Sequential Games: Timing is Everything

In the game of life, timing is everything. Whether it's knowing when to make a move or when to hold back, understanding the concept of sequential games is crucial. Sequential games involve multiple players, each taking turns to make decisions, with each decision affecting the outcome of the game. In this section, we will dive into the intricacies of sequential games, exploring the strategies and tactics that can lead to success.

Basics of Sequential Games

In sequential games, players take turns to make decisions, and the order of these decisions plays a critical role in determining outcomes. Unlike simultaneous games where all players make their decisions simultaneously, sequential games unfold step by step, allowing players to observe the moves of previous players before making their own.

To set the stage for understanding sequential games, let's consider an example involving two players, Alice and Bob. They are playing a game where each has two choices: cooperate (C) or defect (D). The possible outcomes and payoffs are as follows:

Alice / Bob	Cooperate (C)	Defect (D)
Cooperate (C)	(3, 3)	(0, 5)
Defect (D)	(5, 0)	(1, 1)

In this game, the first number in each cell represents Alice's payoff, while the second number represents Bob's payoff. The payoffs are arranged in the format (Alice's payoff, Bob's payoff).

Backwards Induction

To analyze sequential games, we often employ a technique called backwards induction. This approach involves starting from the last player's turn and reasoning backward, anticipating the optimal strategies of each player at every decision point.

In our example, let's consider Bob's decision first. He can either cooperate (C) or defect (D). Assuming Alice will play optimally, Bob knows that if he cooperates, Alice's best response will be to defect to

TWO-PLAYER ZERO-SUM GAMES

maximize her payoff. Thus, Bob's rational choice is to defect as well, resulting in a payoff of (1, 1).

Next, let's analyze Alice's decision. With the knowledge that Bob will choose to defect, Alice must decide whether to cooperate or defect. If she cooperates, her payoff will be 0. However, if she defects, she can increase her payoff to 1. Therefore, Alice's rational choice is to defect, resulting in a payoff of (1, 1) for both players.

By reasoning backward, we have determined the optimal outcome of this sequential game: both Alice and Bob defect, with each receiving a payoff of 1. This process of reasoning backward, considering each player's optimal strategy at every decision point, is the essence of backwards induction.

Timing and Strategic Moves

In sequential games, timing plays a crucial role in determining the outcome. A player must carefully consider the actions of other players and strategically time their move for maximum advantage. Let's consider another example to illustrate the significance of timing in sequential games.

Suppose Alice and Bob are playing a game where each has three options: attack (A), defend (D), or wait (W). The payoffs for each combination are as follows:

Alice / Bob	Attack (A)	Defend (D)	Wait (W)
Attack (A)	(0, 0)	(2, 1)	(-1, -1)
Defend (D)	(1, 2)	(0, 0)	(1, -1)
Wait (W)	(-1, 1)	(-1, 1)	(0, 0)

In this game, both Alice and Bob must decide whether to attack, defend, or wait. The resulting payoffs are given in the format (Alice's payoff, Bob's payoff).

To determine the optimal strategy, players must consider the potential moves of their opponents and the corresponding payoffs. By reasoning backward, we can calculate the optimal strategy for this sequential game.

Let's start with Bob's turn. If Alice attacks, Bob's best response is to defend, resulting in a payoff of (2, 1). If Alice defends, Bob's best response is to attack, leading to a payoff of (1, 2). Finally, if Alice waits,

Bob's best response is to wait as well, resulting in a payoff of (0, 0). In this case, Bob's rational choice is to wait, as it guarantees him a payoff of 0.

Once we determine Bob's optimal strategy, we can move on to Alice's turn. If Bob waits, Alice's best response is to wait as well, ensuring a payoff of (0, 0). If Bob attacks, Alice's best response is to defend, leading to a payoff of (2, 1). Lastly, if Bob defends, Alice's best response is to attack, resulting in a payoff of (1, 2). In this case, Alice's rational choice is to attack.

Analyzing the game using backwards induction, the optimal outcome is for Alice to attack and Bob to wait, resulting in a payoff of (1, 0) for Alice and (0, 0) for Bob.

Real-World Applications

The concept of sequential games finds applications in various real-world scenarios. Consider a negotiation between two companies over a contract. Each company must strategically time its moves, taking into account the other company's actions and responses. By effectively analyzing the sequence of moves and employing backward induction, companies can negotiate more favorable outcomes.

Sequential games also play a role in sports, particularly in games with alternating turns, such as tennis or chess. Players must carefully consider the timing of their moves and anticipate their opponent's strategies to achieve success in these games.

In conclusion, understanding sequential games is essential for strategic decision-making in various fields. By recognizing the importance of timing and employing backwards induction, individuals can better navigate complex decision-making scenarios, leading to more favorable outcomes. So remember, in game theory, timing is everything!

Exercises

1. Consider a sequential game between a buyer and a seller in an online auction. The buyer has two options: place an early bid (E) or wait until the auction is about to end (W). The seller, in response, can either start the bidding low (L) or set a high initial bid (H). The payoffs for each combination are as follows:

Buyer / Seller	Low (L)	High (H)
Early (E)	(10, 5)	(3, 3)
Wait (W)	(2, 2)	(8, 7)

a) Determine the optimal strategy for each player using backwards induction.

b) What is the resulting outcome of this sequential game?

2. Imagine a scenario where two companies, A and B, are competing for a government contract. The contract will be awarded based on the order in which the companies submit their bids. If both companies submit their bids simultaneously, the government will choose the bid with the lowest cost. However, if one company submits its bid before the other, the second company has the opportunity to revise its bid after observing the first bid. The payoffs for each outcome are as follows:

- If A submits first and B revises: (10, 7)

- If A submits first and B does not revise: (5, 5)

- If B submits first and A revises: (7, 10)

- If B submits first and A does not revise: (5, 5)

Assuming both companies aim to maximize their payoffs, determine the optimal strategies for each company and the resulting outcome of this sequential game.

Additional Resources

For further exploration of sequential games and game theory, you may find the following resources helpful:

- Dixit, A., & Nalebuff, B. (2008). *The Art of Strategy: A Game Theorist's Guide to Success in Business and Life*. Norton.

- Osborne, M. J., & Rubinstein, A. (1994). *A Course in Game Theory*. MIT Press.

- Binmore, K. (2007). *Playing for Real: A Text on Game Theory*. Oxford University Press.

These resources offer comprehensive coverage of game theory, including various types of games, strategies, and real-world applications. Happy gaming!

Extensive Form and Perfect Information Games

In game theory, extensive form and perfect information games are two important concepts that help us analyze and understand the strategic interactions in a game. These concepts provide a framework for representing games that involve sequential decision-making and where players have complete and perfect information about the game.

What are Extensive Form Games?

Extensive form games, also known as sequential games, represent a sequence of moves taken by players in a specific order. This representation captures the temporal element of decision-making, allowing us to analyze the game tree and identify the optimal strategies for each player.

To illustrate this, let's consider a classic sequential game: "Rock, Paper, Scissors". In this game, Player 1 chooses either rock, paper, or scissors, and Player 2 chooses one of the same three options. The outcome of the game depends on the choices made by both players simultaneously.

We can represent this game in the extensive form by creating a game tree. The tree starts with Player 1's decision node, followed by Player 2's decision node, and finally ends with the terminal nodes representing the outcomes. Each edge represents a choice made by a player, leading to a new decision node or a terminal node.

TWO-PLAYER ZERO-SUM GAMES

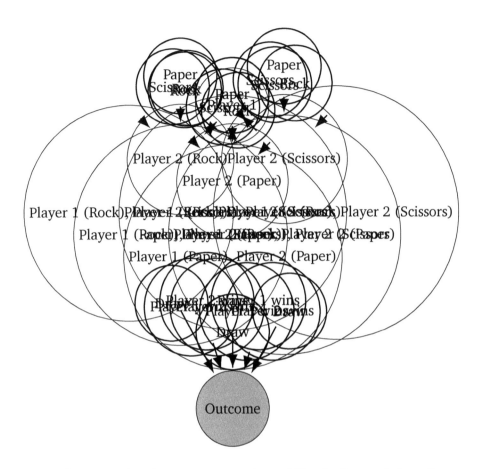

In the extensive form representation of "Rock, Paper, Scissors", we can clearly see the sequence of moves taken by each player and the corresponding outcomes. This representation is useful when analyzing strategies and finding the optimal choices for each player.

What are Perfect Information Games?

Perfect information games, as opposed to imperfect information games, are games where each player has complete information about the actions, payoffs, and strategies of all other players in the game. In other words, there are no hidden or private moves.

A classic example of a perfect information game is chess. In chess, players have full knowledge of the current state of the game, the available moves, and the consequences of those moves. This allows for a thorough analysis of the game and the development of advanced strategies.

To represent perfect information games, we can use the extensive form. Each node in the game tree represents a decision point, and the edges represent the actions available to the players. The terminal nodes represent the final outcomes or payoffs for the players.

Consider the following example:

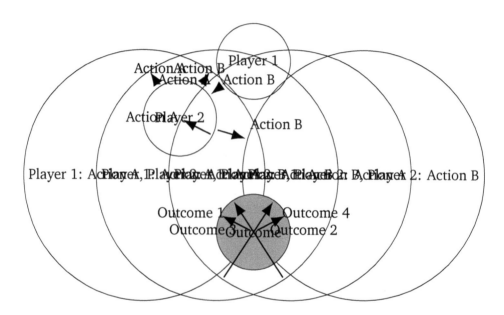

In this example, Player 1 and Player 2 have two possible actions each: Action A and Action B. The outcomes of the game depend on the combined actions of the players. By analyzing the game tree, we can determine the optimal strategies for each player and the resulting payoffs.

Strategies in Extensive Form and Perfect Information Games

In extensive form and perfect information games, players choose strategies based on the information they have at each decision point. A strategy is a complete plan of action that specifies which action to take at each possible decision point in the game.

To understand strategies in these types of games, we can use the concept of backward induction. Starting from the final outcome or terminal node, we work backward through the game tree, determining the optimal actions for each player at each decision point.

For example, let's consider a simple extensive form game where Player 1 can choose between taking Action X or Action Y, and Player 2 can choose between taking Action A or Action B. The payoffs for each outcome are shown in the game tree.

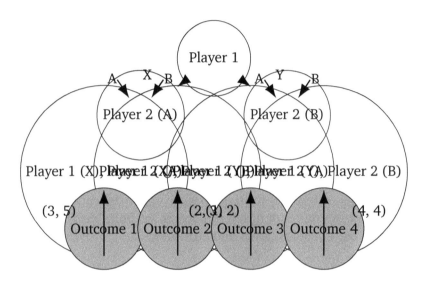

To determine the optimal strategies for each player, we start from the final outcomes and work backward. In this case, we find that Player 2 will choose Action B, and Player 1 will choose Action Y.

By using backward induction, we can solve extensive form games and identify the optimal strategies for each player. This allows us to gain insights into the dynamics of the game and predict the likely outcomes based on players' rational decision-making.

Real-World Applications and Examples

Extensive form and perfect information games have numerous real-world applications in various fields. Let's explore a few examples:

Chess Strategy In the game of chess, players make sequential moves in a perfect information game. Understanding the extensive form representation of chess allows players to analyze the game tree, evaluate different strategies, and make optimal moves to gain an advantage over their opponents.

Negotiation Tactics In negotiations, parties often make sequential offers and counteroffers. By representing negotiation scenarios in the extensive form, we can analyze the possible outcomes, identify strategic moves, and develop effective negotiation tactics.

Poker Analysis Poker is a game that involves sequential decision-making and imperfect information. However, by modeling it as an extensive form game with perfect information, we can take into account the information revealed through betting and use probabilistic analysis to make optimal decisions based on game theory principles.

Supply Chain Management Supply chain management involves decisions made in a sequential manner, where each player in the supply chain can impact the overall outcome. By modeling supply chain interactions as extensive form games, companies can analyze the strategies and incentives of different players, optimize decision-making, and achieve better coordination and efficiency.

Military Operations Military operations often involve sequential decision-making where commanders must plan and execute actions based on the actions of the opposing forces. Analyzing military operations as extensive form games can help strategists understand the dynamics of conflicts, anticipate enemy moves, and devise effective strategies to achieve their objectives.

Summary

Extensive form and perfect information games provide a powerful framework for analyzing and understanding strategic interactions in games. By representing games in the extensive form and considering perfect information settings, we can analyze the game tree, identify optimal strategies, and predict the likely outcomes of the game. These concepts have numerous real-world applications in various fields, from chess and negotiation to supply chain management and military operations. By utilizing these game theory principles, individuals and organizations can make more informed decisions, gain advantages in competitive situations, and achieve better overall outcomes. So get ready to play smart, not hard!

TWO-PLAYER ZERO-SUM GAMES

Game Trees: Visualizing Sequential Games

In the previous sections, we explored various aspects of game theory, including decision-making, types of games, and strategic interactions. Now, let's delve into game trees, a powerful tool for visualizing and analyzing sequential games.

Understanding Game Trees

Game trees are graphical representations that illustrate the sequence of moves and potential outcomes in a sequential game. They allow us to visualize the decision-making process and understand the strategic interactions between players.

A game tree consists of nodes and branches. The nodes represent decision points for each player, and the branches represent the possible choices available at each node. The game tree branches out as the game progresses, creating a tree-like structure.

To construct a game tree, we start with an initial node representing the starting point of the game. From this node, we draw branches to represent the possible actions available to the players. Each subsequent node along a branch represents a decision point, with branches stemming from it representing further choices.

Analyzing Game Trees

Analyzing a game tree involves determining the possible outcomes and optimal strategies for each player. This can be done by working backwards from the terminal nodes (end points of the game) to the initial node, using a technique called backward induction.

At the terminal nodes, we assign payoffs to each player based on the outcome of the game. These payoffs represent the utility or value that each player receives. By assigning payoffs, we can determine the players' preferences and assess the overall outcome of the game.

Starting from the terminal nodes, we work backward through the tree, applying the concept of rationality. At each decision node, players evaluate the potential payoffs of their available choices and select the option that maximizes their expected utility.

Example: The Prisoner's Dilemma

Let's use the famous Prisoner's Dilemma as an example to understand how game trees work. In this game, two prisoners are given the choice to either cooperate with or betray each other. The payoffs for each outcome are as follows:

	Prisoner B Cooperate (C)	Betray (B)
Prisoner A	3, 3 (3 each)	0, 5 (A: 0, B: 5)

Let's construct the game tree for the Prisoner's Dilemma:

```
A: C      3,3     B: C     0,5     A: B              B: B
 O-----------------O----------------O-----------------O
```

In this game tree, Player A is the first to make a decision, represented by the initial node. Player A has two choices: cooperate (C) or betray (B). Similarly, Player B, represented by the second level of nodes, also has two choices.

By analyzing this game tree using backward induction, we can determine the optimal strategies for each player. In this case, betraying each other (choosing the rightmost branches) leads to a higher payoff for both players individually, even though mutual cooperation (choosing the leftmost branches) would yield a higher total payoff.

Real-World Applications

Game trees have extensive real-world applications in various fields. They are commonly used in economics, political science, biology, and computer science.

In economics, game trees are employed to model strategic interactions between firms in industries like oligopoly, where a few large firms dominate the market. Game trees help firms analyze their rivals' potential moves, plan their strategies accordingly, and make optimal decisions.

In political science, game trees are utilized to examine strategic negotiations and decision-making processes in international relations. They help decipher the complex dynamics of conflicts, alliances, and negotiations between countries, aiding policymakers and diplomats in making informed choices.

In biology, game trees offer insights into the evolution of behavior and cooperation between organisms. They help scientists understand how certain strategies, such as cooperation or defection, evolve over time and persist in specific ecological contexts.

In computer science, game trees underpin the design and implementation of intelligent algorithms and systems. By utilizing game-theoretic approaches, researchers develop solutions for problems ranging from artificial intelligence and machine learning to multi-agent systems and algorithmic game theory.

Overall, game trees provide a visual and analytical framework for understanding sequential decision-making and strategic interactions. By mapping out the possible choices and outcomes, they help us navigate complex situations and make rational decisions in various domains.

Exercises

1. Construct a game tree for the following scenario: Two friends, Alice and Bob, are deciding whether to study together for an upcoming exam or to use the time to relax and play video games. The payoffs for each outcome are as follows:

	Bob Study (S)	Relax (R)
Alice	5, 5 (5 each)	0, 8 (A: 0, B: 8)

2. Analyze the game tree from Exercise 1 using backward induction to determine the optimal strategies for Alice and Bob. Discuss the implications of their choices.

3. Research and discuss a real-world application of game trees outside the domains mentioned in this section. How does the use of game trees contribute to understanding and decision-making in that particular field?

Resources

- Osborne, M. J., & Rubinstein, A. (1994). *A Course in Game Theory*. MIT Press.

- Myerson, R. B. (2013). *Game Theory: Analysis of Conflict.* Harvard University Press.

- Binmore, K. G. (2007). *Game Theory: A Very Short Introduction.* Oxford University Press.

- Gibbons, R. (1992). *Game Theory for Applied Economists.* Princeton University Press.

Conclusion

Game trees provide a visual and analytical tool to understand and analyze sequential games. By breaking down decisions and potential outcomes, game trees help us determine optimal strategies, assess payoffs, and explore real-world applications in diverse fields. Understanding game trees equips us with a valuable skillset to navigate strategic interactions and make informed choices in complex situations. So let's dive into the fascinating world of game trees and play smart, not hard!

Solving Sequential Games: Backwards Induction

Understanding Sequential Games

In game theory, sequential games are those in which players make their decisions in a specific order, taking into account the previous decisions made by other players. These games involve a series of moves or actions that occur over time, creating a dynamic and strategic environment.

In sequential games, players have the opportunity to observe and react to the actions of other players before making their own decisions. This interdependence among players makes the analysis of these games more complex than simultaneous games, where players make their decisions at the same time.

To effectively analyze and solve sequential games, a powerful tool called "backwards induction" is often used. Backwards induction involves reasoning backwards from the end of the game to determine the optimal strategies for each player at every stage of the game.

Backwards Induction Process

The backwards induction process begins by considering the terminal stage of the game, where no further moves are possible. From this final stage, players determine the best possible outcome or payoff they can achieve, given the actions of the other players.

Once the optimal outcomes at the final stage are determined, players move one step back in the game and analyze the previous stage. At each stage, players consider all possible actions and their corresponding payoffs, assuming that the other players will make their rational decisions.

Players continue to move backwards and analyze each preceding stage until they reach the initial stage of the game. At the initial stage, players have complete information about the entire game and can determine their optimal strategies based on their analysis of future moves.

An Example: Backwards Induction in a Business Rivalry

Let's understand the concept of backwards induction through an example involving two competing businesses: Company A and Company B. The two companies are considering launching a new product in the market. The success of their product launch depends on their advertising strategy.

The game consists of two stages: stage 1 is the advertising decision, and stage 2 is the product launch. The payoffs for each company at each stage are represented in the following game matrix:

	Advertise (A)	Don't Advertise (D)
Advertise (A)	5, 10	0, 0
Don't Advertise (D)	0, 0	10, 5

In this game, the first player is Company A, and the second player is Company B.

We start the backwards induction process by analyzing stage 2, the product launch stage. We observe that if Company A advertises in stage 1 and Company B doesn't, the payoffs at stage 2 are 5 for Company A and 10 for Company B. On the other hand, if both companies don't advertise, the payoffs are 0 for both companies.

Since the payoffs at stage 2 are higher if both companies don't advertise, we eliminate the strategy of advertising for both players in stage 1. We update the game matrix accordingly:

	Advertise (A)	Don't Advertise (D)
Advertise (A)	- , -	0, 0
Don't Advertise (D)	0, 0	- , -

Now, we move to stage 1 of the game. At this stage, both players don't have any advertising options left. Therefore, we leave these cells blank.

Analyzing the updated game matrix, we can see that there is no decision left at either stage of the game. The outcomes in both stages are blank, indicating that there is no further analysis needed as all possible actions have been eliminated.

Limitations and Caveats

While backwards induction is a powerful tool in solving sequential games, it does have some limitations and caveats.

Firstly, the effectiveness of backwards induction relies on the assumption of perfect rationality and complete information. If players have limited information or their decision-making is influenced by emotions or biases, the outcomes of the game may differ.

Additionally, backwards induction assumes that all players are aware of the rationality of their opponents and can accurately predict each other's moves. In reality, players may make mistakes or have varying levels of strategic thinking, which can affect the predicted outcomes.

Furthermore, the process of backwards induction becomes more complex as the number of stages and players in the game increases. The analysis becomes computationally challenging and may require advanced mathematical tools and techniques.

Conclusion

In conclusion, backwards induction is a powerful technique used to solve sequential games in game theory. By reasoning backwards from the end of the game, players can determine their optimal strategies at each stage, taking into account the actions of other players.

While backwards induction provides valuable insights into strategic decision-making, it is essential to consider the limitations and assumptions associated with its application. Real-world scenarios may involve additional complexities and uncertainties that may affect the predicted outcomes.

Understanding and applying backwards induction can enhance decision-making skills and provide a deeper understanding of how players strategize and interact in sequential games. By mastering this technique, individuals can make smarter choices in various domains, ranging from business and economics to politics and social interactions.

Applications of Two-Player Zero-Sum Games in Economics and Politics

In the previous sections, we explored the fundamentals of game theory and its significance in decision-making and strategic interactions. Now, let's dive into how game theory applies specifically to two-player zero-sum games in the realms of economics and politics. These games are characterized by their competitive nature, where one player's gain is exactly balanced by the other player's loss.

Economic Applications

Game theory has profound implications in economics, helping us understand and predict various economic phenomena. Let's look at some key applications:

1. **Pricing Strategies:** Pricing is a crucial aspect of any business strategy. Two-player zero-sum games allow us to analyze the interplay between competing firms and determine optimal pricing strategies. For example, in the airline industry, two rival airlines engage in a pricing game where they continually adjust their fares to attract customers while maximizing their own profits.

2. **Trade Wars:** International trade often involves strategic decision-making by governments. Two-player zero-sum games help us understand the dynamics of trade wars, where countries impose tariffs and trade barriers to protect their domestic industries. Through game theory analysis, policymakers can anticipate the actions of other nations and design effective strategies for negotiations.

3. Auctions: Auctions are prevalent in various sectors, such as art, real estate, and telecommunications. Game theory helps us design auction mechanisms that ensure fairness, efficiency, and optimal revenue generation. For instance, in a simultaneous ascending-bid auction, multiple bidders compete to acquire a single item, and game theory allows us to analyze the best strategies for both buyers and sellers.

4. Market Competition: Two-player zero-sum games provide insights into market competition and the formation of oligopolies. By studying strategic interactions between firms, economists can explore the consequences of collusive behavior and analyze the effects of different strategies on market outcomes.

5. Resource Allocation: Game theory aids in understanding the allocation of scarce resources. In competitive situations where resources need to be divided among players, such as water rights or fishing quotas, game theory provides a framework to determine equitable and efficient allocation mechanisms.

Political Applications

Game theory can also shed light on complex political scenarios and help us analyze the behavior of political actors. Here are some examples:

1. Election Campaigns: Elections involve strategic decision-making by political candidates. Two-player zero-sum games enable us to model election campaigns and analyze the strategies employed by candidates to maximize their chances of winning. Factors such as fundraising, policy positions, and targeting specific voter demographics can be examined through game theory.

2. International Negotiations: Game theory has significant applications in international relations and conflict resolution. By modeling interactions between nations as two-player zero-sum games, we can study how different strategies and actions influence the outcomes of negotiations, treaties, and diplomatic efforts.

3. Legislative Bargaining: In parliamentary systems, legislative decision-making often requires bargaining between political parties. Game theory helps us analyze coalitional behavior, the formation of governing coalitions, and the distribution of power within these coalitions.

4. Political Campaign Finance: Political campaigns rely on financial contributions, and game theory can contribute to our understanding of fundraising strategies and dynamics. By modeling interactions between political donors and candidates as two-player zero-sum games, we can explore the impact of campaign finance regulations and the strategies employed by candidates to secure funding.

5. Arms Races: Game theory provides insights into the dynamics of arms races and the strategies employed by countries to gain a military advantage. By modeling these interactions as two-player zero-sum games, we can study the potential for escalation, the impact of arms control agreements, and the role of deterrent strategies.

In conclusion, two-player zero-sum games find diverse applications in economics and politics. They allow us to analyze competitive scenarios, understand strategic decision-making, and predict outcomes in a wide range of real-world situations. By employing game theory, we gain valuable insights into the behavior of economic agents and political actors, helping us make informed decisions and design effective strategies. Game theory is truly a powerful tool in the arsenal of economists and political scientists alike.

To further explore these applications, I recommend the following resources:

- *Game Theory for Applied Economists* by Robert Gibbons.

- *The Strategy of Conflict* by Thomas C. Schelling.

- *An Introduction to Game Theory* by Martin J. Osborne and Ariel Rubinstein.

Now, let's test your understanding with a problem!

Problem: Two mobile phone companies, XCell and TeleMatch, are competing for market share in a city. They can either set high prices (H) or low prices (L). The following payoff matrix shows the profits for each company, where positive profits represent gains and negative profits represent losses (in millions of dollars):

	XCell (H)	XCell (L)
TeleMatch (H)	5, 5	-1, 10
TeleMatch (L)	10, -1	2, 2

Assuming both companies want to maximize their profits, analyze the strategies they should adopt and determine the Nash equilibrium of the game.

Solution:

To find the Nash equilibrium, we need to identify each company's best response strategy given their competitor's strategy. Let's consider each company's options:

For XCell:

- If TeleMatch sets high prices (H), XCell's profits are maximized by also setting high prices (5).

- If TeleMatch sets low prices (L), XCell's profits are maximized by setting low prices (2).

For TeleMatch:

- If XCell sets high prices (H), TeleMatch's profits are maximized by setting low prices (10).

- If XCell sets low prices (L), TeleMatch's profits are maximized by setting low prices (-1).

Considering these options, both companies' best response strategy is to set low prices (L). Thus, (L, L) is the Nash equilibrium of the game.

This scenario illustrates a price competition where both companies benefit from setting low prices rather than engaging in a price war. Such insights can help companies devise optimal strategies to maximize their profits in competitive markets.

Remember, game theory allows us to unravel the complex dynamics of economic and political interactions. By understanding strategic decision-making, we can navigate a world where smart plays matter more than hard work alone.

Now that you've explored the applications of two-player zero-sum games in economics and politics, you're ready to dive deeper into the fascinating world of game theory. In the next section, we'll explore cooperative games and the possibilities they bring.

Cooperative Games

Characteristics of Cooperative Games

Cooperative games are a special class of games where players form coalitions and work together towards achieving a common goal. Unlike competitive games, where players solely focus on their individual outcomes, cooperative games introduce the concept of collaboration and joint decision-making. In this section, we will explore the key characteristics of cooperative games and understand how they differ from other types of games.

Shared Goals and Mutual Gain

The defining feature of cooperative games is the presence of shared goals and the potential for mutual gain. In these games, players recognize that by working together, they can achieve outcomes that are superior to what they could achieve individually. Cooperation allows players to combine their skills, resources, and efforts, leading to a collective advantage.

For example, consider a group of friends starting a business together. Each individual brings their unique expertise and contributes to the success of the venture. By cooperating, they can pool their resources, share the workload, and maximize their chances of achieving a profitable outcome.

Coalitions and Team Formation

In cooperative games, players form coalitions or teams to pursue their shared objectives. A coalition is a group of players who agree to work together and make joint decisions. These coalitions can range in size and composition depending on the game and the nature of the interactions.

The formation of coalitions introduces the element of strategy, as players must consider who to ally with and how to build a team that is efficient, effective, and reliable. The dynamics of coalition formation can significantly impact the outcomes of cooperative games.

Negotiation and Communication

Cooperative games often require negotiation and communication among players. Since cooperation involves making joint decisions, players need to discuss strategies, allocate resources, and coordinate their actions. Effective communication is crucial for building trust, resolving conflicts, and maximizing the collective benefits.

Negotiation plays a central role in cooperative games, as players bargain over the distribution of rewards and the division of responsibilities. Successful negotiation relies on skills such as persuasion, compromise, and strategic thinking. It is essential to find mutually acceptable agreements that motivate players to cooperate and align their interests.

Fairness and Equity

Cooperative games place considerable emphasis on fairness and equity. Since players collaborate towards a common goal, it is important to ensure that the outcomes are distributed in a just and equitable manner. Fairness ensures that each player perceives the cooperative endeavor as worthwhile and meaningful.

Various fairness principles have been proposed in cooperative game theory. One widely used concept is the Shapley value, which attributes the contributions of each player to the overall value created by the coalition. The Shapley value provides a fair and efficient solution for allocating rewards among the members of a cooperative game.

Challenges in Cooperation

Cooperative games are not without their challenges. The inherent nature of cooperation introduces potential complications that can impact the outcomes. Some common challenges in cooperative games include:

1. Free-Riding: The phenomenon of free-riding occurs when some players benefit from the efforts of others without contributing proportionately. It can lead to a breakdown in cooperation and hinder the achievement of goals.

2. Communication Costs: Effective communication and coordination among players can be challenging, especially in large-scale cooperative games. Coordinating actions, sharing information, and building trust may require significant time and effort.

3. Incomplete Information: In some cooperative games, players may not have complete knowledge about the preferences, actions, or

abilities of others. Limited information can hinder decision-making and coordination among players.

4. Incentive Misalignment: Aligning the incentives of all players in a cooperative game is crucial for maintaining cooperation. Conflicting interests and divergent goals can undermine cooperation and lead to suboptimal outcomes.

Addressing these challenges and finding solutions to promote cooperation is a major focus in cooperative game theory.

Real-World Applications

Cooperative game theory has numerous real-world applications across various domains. Some notable examples include:

- International collaborations: Cooperative game theory aids in understanding how nations can cooperate to tackle global challenges such as climate change, trade agreements, or security alliances.

- Joint ventures and partnerships: Businesses often form partnerships to combine resources, share risks, and exploit new opportunities. Cooperative game theory provides insights into forming stable and mutually beneficial alliances.

- Teamwork and project management: Cooperative game theory concepts are applicable in managing teams and projects where individuals must collaborate to achieve common objectives.

- Resource allocation: Cooperative game theory can assist in fair allocation of limited resources in situations like water distribution, public goods provision, or disaster relief efforts.

Understanding the characteristics of cooperative games helps us navigate the complexities of cooperative decision-making and collaboration. By analyzing and studying these games, we can develop strategies and frameworks that promote successful cooperation and the realization of mutual gains.

Exercises

1. Consider a group of students working on a group project for a course. Identify the challenges they might encounter in cooperating effectively. Suggest strategies to overcome these challenges.

2. Imagine a scenario where two companies are considering a merger. Discuss the characteristics of a successful cooperative game in this context. How would you ensure fairness and equity in the distribution of benefits?

3. Research and analyze a real-world case study where cooperation among nations was crucial. Identify the cooperative strategies employed and evaluate their effectiveness in achieving the desired outcomes.

4. Investigate the concept of coalition formation in politics. How do political parties form alliances and coalitions during elections or in the formation of government? Discuss the benefits and challenges of such alliances.

Resources

- Books:

    - "Cooperative Game Theory and Applications" by Myrna H. Wooders
    - "Game Theory: Analysis of Conflict" by Roger B. Myerson
    - "The Shapley Value: Essays in Honor of Lloyd S. Shapley" edited by Alvin E. Roth and Marilda A. Oliveira Sotomayor

- Websites:

    - Stanford Encyclopedia of Philosophy - Game Theory: plato.stanford.edu/entries/game-theory
    - Game Theory Society: www.gametheorysociety.org
    - Institute for Operations Research and the Management Sciences (INFORMS): www.informs.org

Takeaways

- Cooperative games involve shared goals and mutual gain, emphasizing collaboration and joint decision-making.

- Players form coalitions and work together to achieve their shared objectives, requiring negotiation and effective communication.

- Fairness and equity play a crucial role in cooperative games, with various principles for distributing rewards among the players.

- Free-riding, communication costs, incomplete information, and incentive misalignment are common challenges in cooperative games.

- Cooperative game theory has applications in international collaborations, joint ventures, project management, and resource allocation.

Remember, in cooperative games, teamwork makes the dream work (but only if everyone puts in the effort)!

Shapley Value: Dividing the Pie Fairly

In cooperative games, where players can form coalitions and work together to achieve common goals, the question of how to distribute the gains fairly becomes crucial. The Shapley value provides a solution by assigning a value to each player, reflecting their contribution and importance in the game.

The concept of the Shapley value was introduced by Lloyd Shapley in the 1950s and has since become a fundamental tool in cooperative game theory. It is based on the idea that each player's worth is determined by the marginal contributions they make when joining different coalitions.

To understand the Shapley value, let's consider a simple example. Imagine a group of friends deciding to start a business together. Each friend brings a unique set of skills and resources to the table, and they want to divide the profits fairly.

The Shapley value offers a fair solution by considering all possible permutations of the players' arrival order. It calculates the average contribution of each player across all possible coalitions, taking into account the varying sizes of the coalitions.

To calculate the Shapley value for a player, we sum their marginal contributions when they join different coalitions, and then divide that sum by the total number of possible permutations.

Let's illustrate this with an example:

Imagine a coalition of four friends, Alex, Ben, Claire, and Dana, starting a karaoke business. The profits from the business will be divided amongst the friends, but they need to determine a fair allocation based on their contributions.

Here's a breakdown of their potential earnings with different permutations of the friends arriving:

- Alex, Ben, Claire, Dana: 1000
- Alex, Claire, Ben, Dana: 900
- Ben, Alex, Claire, Dana: 800
- Ben, Claire, Alex, Dana: 1000
- Claire, Alex, Ben, Dana: 800
- Claire, Ben, Alex, Dana: 900

To calculate the Shapley value for each friend, we consider all possible permutations and calculate their marginal contributions. Let's see the step-by-step calculations for each friend:

Calculating Shapley Value for Alex

- Alex's marginal contribution when joining as the first player: 1000
- Alex's marginal contribution when joining as the second player: $900 - 800 = 100$
- Alex's marginal contribution when joining as the third player: $800 - 800 = 0$
- Alex's marginal contribution when joining as the fourth player: 0

To calculate the average contribution, we sum up the marginal contributions: $1000 + 100 + 0 + 0 = 1100$. Since there are 4! total permutations of the players, the Shapley value for Alex is $1100/24 = 45.83$.

Similarly, we can calculate the Shapley value for the other friends:

Calculating Shapley Value for Ben

- Ben's marginal contribution when joining as the first player: 0
- Ben's marginal contribution when joining as the second player: $1000 - 800 = 200$
- Ben's marginal contribution when joining as the third player: $1000 - 900 = 100$
- Ben's marginal contribution when joining as the fourth player: 0

The sum of Ben's marginal contributions is $0 + 200 + 100 + 0 = 300$. Therefore, Ben's Shapley value is $300/24 = 12.5$.

Calculating Shapley Value for Claire

- Claire's marginal contribution when joining as the first player: 0
- Claire's marginal contribution when joining as the second player: $900 - 800 = 100$
- Claire's marginal contribution when joining as the third player: $900 - 800 = 100$
- Claire's marginal contribution when joining as the fourth player: 0

The sum of Claire's marginal contributions is $0 + 100 + 100 + 0 = 200$. Therefore, Claire's Shapley value is $200/24 = 8.33$.

Calculating Shapley Value for Dana

- Dana's marginal contribution when joining as the first player: 0
- Dana's marginal contribution when joining as the second player: 800
- Dana's marginal contribution when joining as the third player: 900
- Dana's marginal contribution when joining as the fourth player: 1000

The sum of Dana's marginal contributions is $0 + 800 + 900 + 1000 = 2700$. Therefore, Dana's Shapley value is $2700/24 = 112.5$.

Now that we have calculated the Shapley value for each friend, we can determine the fair allocation of profits. In this case, Alex would receive approximately 45.83, Ben would receive 12.5, Claire would receive 8.33, and Dana would receive 112.5.

The Shapley value guarantees that each player is rewarded based on their individual contribution and the order of their arrival. It provides a fair and unbiased solution for dividing the profits in cooperative games.

The Shapley value has wide-ranging applications in various fields, including economics, politics, and even fair resource allocation in network routing. Its ability to allocate rewards fairly makes it a valuable tool for decision-making when multiple parties are involved.

Further Reading

If you want to dive deeper into the concept of the Shapley value and its applications, here are some resources to explore:

- Roth, A. E. (1988). *The Shapley Value: Essays in Honor of Lloyd S. Shapley*.

- Moulin, H. (1988). *Axioms of Cooperative Decision Making*.

- Myerson, R. B. (1997). *Game Theory: Analysis of Conflict*.

These resources delve into more advanced topics and offer a broader understanding of cooperative game theory and the various methods for fair resource allocation.

Exercise

Imagine a scenario where four countries are working together to combat climate change. Each country has a different carbon emissions reduction target, and they need to decide how to distribute the financial burden of achieving their goals fairly. Use the concept of the Shapley value to calculate the fair allocation of costs for each country based on their emission reduction efforts.

Hint: Consider the marginal contributions of each country when forming different coalitions and calculate their overall Shapley values.

COOPERATIVE GAMES

Take time to think about the potential challenges and ethical considerations involved in this exercise. How might the Shapley value help in addressing the fairness of burden sharing among countries in combating global issues like climate change?

Remember, game theory offers a range of tools and concepts that can be applied to real-world problems, allowing us to make more informed decisions in complex situations. The Shapley value is just one of the many powerful tools in cooperative game theory that helps us strive for fairness and equity.

Core and Stability in Cooperative Games

In cooperative game theory, the core is a fundamental concept that captures a notion of stability in the allocation of payoffs among players in a coalition. It ensures that no subset of players has an incentive to break away and form a new coalition, as they would be worse off outside of the core.

Defining the Core

Formally, the core of a cooperative game is defined as the set of payoff allocations that satisfies two key properties: feasibility and stability.

1. **Feasibility**: For an allocation to belong to the core, it must distribute the total available payoff of the grand coalition (all players) among the players without exceeding their individual contribution to the coalition. In other words, no player can receive more than what they contribute.

2. **Stability**: The core ensures that no coalition of players has an incentive to deviate and form a new coalition with a different payoff allocation. This means that no group of players can collectively improve their situation by leaving the current coalition and forming a new one. The core provides stability by preventing any profitable deviations.

The core captures the idea of a self-enforcing cooperative solution. It represents a collective agreement that all players find acceptable, as no subset of players can improve their payoffs by breaking away.

Shapley Value and Core Allocation

One way to allocate the payoffs in the core is by using the Shapley value. The Shapley value is a fair allocation that takes into account the contributions of individual players to various coalitions.

The Shapley value for a player in a cooperative game is defined as the average marginal contribution they bring to all possible coalitions. It calculates the expected added value that a player brings to a coalition, considering all possible orders in which players can join.

$$SV(i) = \sum_{S \subseteq N \setminus \{i\}} \frac{|S|!(n - |S| - 1)!\,(v(S \cup \{i\}) - v(S))}{n!}$$

where: - $SV(i)$ is the Shapley value of player i, - N is the set of all players, - $v(S)$ is the worth of coalition S, and - n is the total number of players.

The Shapley value provides a fair and stable allocation, as it considers the contributions of each player to various coalitions, giving them a reasonable share of the total payoff.

Core Stability and the Core Cover

While the core guarantees stability, it is not always non-empty. In some games, the core may be empty, indicating that no payoff allocation satisfies both feasibility and stability.

To address this, the concept of the core cover was introduced. The core cover is a larger set that includes the core and allows for additional allocations that may not satisfy stability but are still considered acceptable.

The core cover captures the idea of a weaker notion of stability, where deviations may be permitted if they provide some advantage but still ensure that no player can unilaterally improve their situation.

Real-World Applications

The core and its related concepts find relevance in various real-world scenarios.

In business settings, the core and core cover help in fair profit distribution among stakeholders and provide a framework for analyzing cooperative agreements and partnerships.

In political coalitions, the core ensures the stability of joint decision-making processes and prevents individual parties from breaking off to form new alliances for personal gain.

Furthermore, the core and its extensions have been applied in climate change negotiations, resource allocation in organizations or institutions, and even in analyzing power dynamics in social networks.

Conclusion

Understanding the core and stability in cooperative games is essential for analyzing the fairness and stability of coalitional structures. The core provides a solution concept that guarantees a stable and feasible allocation of payoffs among players, while the core cover allows for additional allocations that may not satisfy full stability.

Cooperative game theory offers valuable insights into decision-making processes and can help individuals and organizations develop strategies to achieve fair and stable outcomes in a wide range of contexts. The core and its related concepts provide a foundation for understanding and analyzing cooperation and the distribution of benefits in various scenarios.

Bargaining and Negotiation in Cooperative Games

In cooperative games, players have the opportunity to form coalitions and work together to achieve common goals. However, when it comes to dividing the rewards of cooperation, there is often a need for bargaining and negotiation. This section explores the strategies and techniques involved in bargaining and negotiation within the context of cooperative games.

Understanding Bargaining in Cooperative Games

Bargaining is the process of discussing and reaching an agreement on how to allocate the outcomes of cooperation. It involves negotiating the distribution of payoffs among the players in a way that is perceived as fair and satisfactory by all parties involved. The challenge lies in finding a solution that satisfies everyone's preferences and avoids conflicts.

Negotiation Strategies in Cooperative Games

Negotiation in cooperative games can take many forms, depending on the specific game and the players involved. Here are some commonly used strategies:

- **Compromise**: This strategy involves each player making concessions and finding a middle ground that is acceptable to all. It requires a willingness to give up certain benefits in exchange for others.

- **Win-win approach**: In this strategy, the focus is on maximizing the joint benefits for all players. It involves cooperative problem-solving and brainstorming to find creative solutions that satisfy everyone's interests.

- **Tit-for-tat**: This strategy is based on reciprocation, where players take turns making offers and responding to each other's offers. It promotes fairness and encourages cooperation by rewarding cooperation and punishing defection.

- **Bullying**: While not an ideal strategy, some players may use aggressive tactics to intimidate others and gain an advantage in negotiations. This can lead to suboptimal outcomes and damage long-term relationships.

Negotiation Techniques in Cooperative Games

Effective negotiation requires certain techniques to ensure a smooth and successful process. Here are some techniques commonly employed:

- **Active listening**: Listening actively to the other party's concerns and perspective is crucial for understanding their needs and finding common ground.

- **Mutual respect**: Respecting the other party's dignity and treating them with fairness and integrity fosters trust and encourages a cooperative atmosphere.

- **Exploring alternatives**: Generating multiple options and exploring different scenarios can help find creative solutions that meet everyone's needs.

- **Analyzing alternatives**: Evaluating the pros and cons of different options allows for informed decision-making and helps in negotiating the best outcome.

- **Maintaining communication**: Clear and open communication throughout the negotiation process is essential for establishing trust and understanding.

The Nash Bargaining Solution

In 1950, mathematician John Nash introduced the concept of the Nash Bargaining Solution, which provides a theoretical framework for bargaining in cooperative games. The solution is based on the idea that a fair outcome is one where no player can be made better off without making another player worse off.

The Nash Bargaining Solution uses a bargaining model to calculate the outcome that maximizes the product of each player's share of the cooperative surplus. The cooperative surplus represents the joint benefit that can be achieved from cooperation.

Mathematically, the Nash Bargaining Solution involves maximizing the product of the players' payoffs subject to certain constraints. This solution provides a fair and efficient outcome that is acceptable to all players.

Real-World Applications

Bargaining and negotiation techniques are widely applicable in various real-world scenarios. Here are some examples:

- **Business negotiations**: In business settings, bargaining and negotiation are common during contract discussions, mergers and acquisitions, and pricing negotiations.

- **Labor negotiations**: Labor unions negotiate with employers to determine wages, working conditions, and benefits for their members. Bargaining techniques are crucial in reaching agreements that are satisfactory to both parties.

- **International relations**: Diplomatic negotiations between countries involve bargaining and negotiation to resolve conflicts, establish treaties, and reach agreements on various issues such as trade, security, and climate change.

- **Legal settlements**: In legal disputes, parties often engage in negotiation to reach a settlement without going to trial. Bargaining techniques help both sides find a mutually acceptable resolution.

- **Collective decision-making**: In group settings, such as community organizations or government committees, bargaining and negotiation play a role in reaching consensus, making joint decisions, and allocating resources.

Exercise

Consider a scenario where two companies, Alpha Corporation and Beta Industries, are negotiating a joint venture. The companies must decide how to divide the profits generated from the venture. Alpha Corporation believes that they should receive 60% of the profits, while Beta Industries argues for a 50% share.

Using the Nash Bargaining Solution, calculate a fair and efficient allocation of profits that satisfies both companies' preferences. Discuss the implications of the outcome and potential strategies that could be employed to reach a compromise.

Additional Resources

If you're interested in learning more about bargaining and negotiation in cooperative games, the following resources provide further insights:

1. Dixit, A., & Nalebuff, B. (2008). *The Art of Strategy: A Game Theorist's Guide to Success in Business and Life*. W. W. Norton & Company.

2. Raiffa, H. (1982). *The Art and Science of Negotiation*. Belknap Press.

3. Rubin, J. Z., Pruitt, D. G., & Kim, S. H. (2009). *Social Conflict: Escalation, Stalemate, and Settlement*. McGraw Hill Education.

4. Shell, G. R. (2006). *Bargaining for Advantage: Negotiation Strategies for Reasonable People*. Penguin.

Remember, bargaining and negotiation are dynamic processes that require practice and patience. By understanding the strategies, techniques, and theoretical foundations discussed in this section, you will be better equipped to navigate and succeed in cooperative game settings.

Transferable Utility and Non-Transferable Utility Games

In the world of game theory, we encounter various types of games that involve different forms of utility. Utility refers to the satisfaction, value, or benefit that individuals derive from particular outcomes or choices. In this section, we will delve into the concepts of transferable utility (TU) and non-transferable utility (NTU) games, which represent two distinct scenarios in which utility is allocated among players.

Transferable Utility Games

In transferable utility games, players have the ability to transfer their utility to others. This means that the value assigned to a particular outcome or choice is not fixed and can be transferred or shared among the players involved. Transferable utility games are often characterized by negotiations, coalitions, and the division of resources.

One of the most commonly used tools in analyzing transferable utility games is the characteristic function. The characteristic function assigns a value to each possible coalition that can be formed among the players. Let's consider a simple example to understand this concept:

Example 3.2.5.1: Suppose there are three friends, Alice, Bob, and Carol, who want to share a pizza. Each friend has a different valuation for the pizza, which they are willing to transfer to others. The characteristic function for this game could be represented as follows:

$$v(S) = \begin{cases} 8 & \text{if } S = \{\text{Alice}\} \\ 6 & \text{if } S = \{\text{Bob}\} \\ 7 & \text{if } S = \{\text{Carol}\} \\ 12 & \text{if } S = \{\text{Alice}, \text{Bob}\} \\ 14 & \text{if } S = \{\text{Alice}, \text{Carol}\} \\ 10 & \text{if } S = \{\text{Bob}, \text{Carol}\} \\ 20 & \text{if } S = \{\text{Alice}, \text{Bob}, \text{Carol}\} \\ 0 & \text{otherwise} \end{cases}$$

In this characteristic function, each singleton coalition (consisting of only one person) has a specific value, while the larger coalitions have different values reflecting the combined preferences of the players involved.

Now, the question is: How should the pizza be divided among the friends to achieve a fair outcome? To address this, we employ the concept of solution concepts, such as the Shapley value and the core, which help us find reasonable allocations.

The *Shapley value* is a method used to distribute the total value generated by the coalition among the players. It takes into account the contribution of each player in different coalitions and calculates a fair share for each individual based on their marginal contribution. Mathematically, the Shapley value can be defined as:

$$\phi_i(v) = \frac{1}{n!} \sum_{\sigma \in S} v(\sigma \cup \{i\}) - v(\sigma)$$

where $\phi_i(v)$ represents the Shapley value of player i, n is the total number of players, $v(\sigma \cup \{i\})$ is the value of the coalition σ with player i added, and $v(\sigma)$ is the value of the coalition σ without player i.

To calculate the Shapley value for our pizza-sharing game, we need to consider all possible orderings of the players and compute their marginal contributions. Let's calculate the Shapley values for Alice, Bob, and Carol:

$$\phi_{\text{Alice}}(v) = \frac{1}{3!}\left[v(\{\text{Alice}\}) + (v(\{\text{Alice}, \text{Bob}\}) - v(\{\text{Alice}\})) + (v(\{\text{Alice}, \text{Carol}\}) -\right.$$
$$= \frac{1}{6}[8 + (12 - 8) + (14 - 8) + (20 - 12)]$$
$$= \frac{1}{6} \times 34 = \frac{17}{3}$$

Similarly, we can calculate the Shapley values for Bob and Carol:

$$\phi_{\text{Bob}}(v) = \frac{1}{6} \times 26 = \frac{13}{3}$$
$$\phi_{\text{Carol}}(v) = \frac{1}{6} \times 30 = 5$$

The Shapley values tell us the fair share that each friend should receive. In this case, Alice should get $\frac{17}{3}$ slices, Bob should get $\frac{13}{3}$ slices, and Carol should get 5 slices.

Another important solution concept in transferable utility games is the *core*. The core is a set of allocations for which no subgroup of players has an incentive to deviate and form a separate coalition to gain more utility. In other words, it represents stable allocations where no one can increase their share without disrupting the cooperation of the group.

To determine if an allocation is in the core, we need to check if the sum of utilities allocated to each player in the allocation is at least as much as the value they can obtain by forming a separate coalition. Let's consider the allocation where Alice gets 5 slices, Bob gets 4 slices, and Carol gets 3 slices. Is this allocation in the core?

To answer this, we need to check if any coalition of two players can deviate and form a separate coalition to gain more utility. Let's consider the coalition of Alice and Bob. If they deviate and form a separate coalition, they can get a total of 9 slices (5 for Alice and 4 for Bob) compared to the 9 slices they are already getting in the current allocation. Since they have no incentive to deviate, the allocation is in the core.

Transferable utility games offer numerous real-world applications. For example, in a business setting, transferable utility games can be used to analyze negotiation scenarios, profit sharing among partners, or resource allocation in project management. Understanding how utility can be transferred and shared among players provides valuable insights into cooperative decision-making processes.

Non-Transferable Utility Games

In contrast to transferable utility games, non-transferable utility (NTU) games involve situations where players cannot transfer or share their utility. Each player's utility in an NTU game is unique and cannot be divided or combined with others. These types of games are often characterized by non-divisible goods or resources.

A common example of an NTU game is an auction, where players bid on an item and the highest bidder wins the item. In this case, each player's utility is determined solely by whether they win the auction or not, and no utility transfer takes place.

In NTU games, the focus is on finding efficient allocations and ensuring fairness. One way to determine efficient outcomes is to analyze dominant strategies and Nash equilibria. A *dominant strategy* is a strategy that yields the highest payoff for a player, regardless of the strategies chosen by other players. On the other hand, a *Nash equilibrium* is a set of strategies where no player has an incentive to deviate from their chosen strategy, given the strategies chosen by others.

Let's consider an example to illustrate the concept of dominant strategies and Nash equilibrium in an NTU game:

Example 3.2.5.2: Suppose Alice and Bob are bidding on a rare collectible comic book. They each have a valuation for the comic book and decide to submit sealed bids. The highest bidder wins the comic book, but only pays the amount of the second-highest bid. Here are their valuations:

Alice's valuation: $200 Bob's valuation: $150

To analyze this auction, we need to consider the possible strategies and payoffs for Alice and Bob. Let's examine the scenario where Alice bids $180 and Bob bids $160:

- If Alice's bid ($180) is higher than Bob's bid ($160), Alice wins the comic book and pays $160. Her payoff is $200 (her valuation) minus $160 (the actual payment), resulting in a net gain of $40.

- If Bob's bid ($160) is higher than Alice's bid ($180), Bob wins the comic book and pays $180. His payoff is $150 (his valuation) minus $180 (the actual payment), resulting in a net loss of $30.

From this analysis, we can observe that Alice's dominant strategy is to bid $180, as it guarantees her a positive payoff regardless of Bob's bid. Bob, on the other hand, does not have a dominant strategy, as his payoff depends on Alice's bid. However, we can determine a Nash equilibrium by considering the best response of each player to the strategy of the other player. In this case, Bob's best response is to bid $160 if Alice bids $180.

The concept of NTU games extends beyond auctions and can be applied to various scenarios, such as job assignments, resource allocation, or matching problems. Understanding the unique utility considerations in these games allows us to design efficient mechanisms that maximize overall social welfare.

Conclusion

Transferable utility and non-transferable utility games provide valuable frameworks for analyzing different types of strategic interactions. Transferable utility games involve the transfer or sharing of utility among players, allowing for negotiations and coalitions. On the other hand, non-transferable utility games focus on unique utilities and efficient allocations. By studying these concepts and employing solution concepts such as the Shapley value, the core, dominant strategies, and Nash equilibria, we can gain insights into various real-world scenarios, ranging from business negotiations to resource allocation in social systems.

Game theory continues to evolve and find applications in diverse fields, from social networks to evolutionary biology, financial markets, and artificial intelligence. As we progress through this book, we will explore these applications and delve deeper into the captivating world of strategic decision-making. So, buckle up and get ready to play smart, not hard!

Coalitions and Power in Cooperative Games

In cooperative games, players often form coalitions to achieve common goals or maximize their individual benefits. The concept of coalitions is central to understanding power dynamics and decision-making in these games. In this section, we will explore the formation of coalitions, the distribution of power within them, and how they affect the outcomes of cooperative games.

Forming Coalitions

Coalitions are formed when two or more players join forces to collectively increase their chances of achieving desirable outcomes. The formation of coalitions depends on a variety of factors, including shared interests, complementary skills, and the ability to cooperate effectively. Players may choose to form coalitions based on strategic considerations, such as securing resources or countering the power of other players.

To illustrate the importance of coalitions, let's consider a real-life example: a business partnership. Two entrepreneurs may decide to join forces and form a coalition to start a new venture. By pooling their resources, skills, and networks, they increase their chances of success compared to operating individually. The formation of coalitions allows for the sharing of risks, resources, and responsibilities, enabling players to achieve better outcomes collectively than they could on their own.

Power in Coalitions

Once coalitions are formed, the allocation of power becomes a key factor in decision-making and resource distribution. Power in coalitions determines the influence each player has over the outcomes and the ability to shape the group's decisions. Power can manifest in various forms, such as control over resources, expertise, or the ability to attract and retain members in the coalition.

One widely used measure to quantify power within coalitions is the Shapley value. The Shapley value assigns a numerical representation of a player's power based on their marginal contribution to different possible coalitions. It considers all possible permutations of coalition formation and calculates each player's average contribution across them. The Shapley value provides a fair and equitable way to allocate power in coalitions, acknowledging the distinct contributions of each player.

Consider a scenario where a group of friends decides to form a coalition to plan a weekend getaway. The person with excellent organizational skills, a large network of contacts, and a reputation for planning successful trips may hold more power in the coalition. Their expertise and ability to attract other members through their connections can significantly impact the decision-making process within the coalition.

Bargaining Power

In addition to power within coalitions, bargaining power also plays a crucial role in cooperative games. Bargaining power refers to a player's ability to negotiate favorable outcomes or influence the distribution of resources. Players with higher bargaining power can secure better deals, allocate resources in their favor, or shape the decision-making process to align with their interests.

Bargaining power is influenced by a range of factors, including the perceived value of a player's contributions, alternative options available, and the level of competition within the coalition. A player's position within the coalition can also affect their bargaining power, as influential players may be able to strategically leverage their position to their advantage.

To understand the concept of bargaining power, let's consider a situation where a group of colleagues is negotiating the allocation of bonuses in a company. A high-performing employee with critical skills and expertise may have more bargaining power compared to their colleagues. They can leverage their valuable contributions to secure a higher bonus or negotiate for other favorable terms.

Examples and Applications

Coalitions and power dynamics are ubiquitous in various fields, including business, politics, and international relations. Understanding these concepts can help us analyze and predict outcomes in real-world scenarios.

In business, alliances and partnerships between companies can be seen as coalitions formed to achieve common goals, such as expanding market reach or developing new products. The distribution of power within these coalitions can significantly impact decision-making, resource allocation, and the overall success of the partnership.

In politics, the formation and dissolution of coalitions are essential elements of governance. Political parties often join forces to gain a majority in parliament or pass legislation. The allocation of power within these coalitions determines the influence each party has over policymaking and government decision-making.

In international relations, coalitions play a crucial role in shaping global alliances and resolving conflicts. Countries form coalitions to enhance their strategic position, exert influence, or address common challenges. Understanding power dynamics within these coalitions is fundamental in analyzing negotiations, conflicts, and policy outcomes.

Real-Life Challenge: Climate Change Coalition

Consider a real-life challenge: forming a coalition to address climate change. In this scenario, various stakeholders, including governments, environmental organizations, and businesses, need to come together and cooperate to mitigate the effects of climate change.

Your task is to analyze the key players involved, their potential power within the coalition, and how bargaining power may affect decision-making and resource allocation. Consider the challenges and opportunities of forming a coalition to address climate change and propose strategies to navigate the power dynamics and achieve common goals.

This real-life challenge provides an opportunity to understand the complexities involved in coalition formation and the significance of power dynamics in driving collective action. By applying game theory principles, such as coalitions and power allocation, we can gain insights into potential strategies for tackling pressing global issues.

Further Reading and Resources

If you are interested in diving deeper into the topic of coalitions and power in cooperative games, the following resources provide valuable insights:

- Bueno de Mesquita, B., Smith, A., Siverson, R., & Morrow, J. (2003). *The Logic of Political Survival*. The MIT Press.

- Osborne, M. J., & Rubinstein, A. (1994). *A Course in Game Theory*. The MIT Press.

- Myerson, R. B. (1997). *Game Theory: Analysis of Conflict*. Harvard University Press.

These resources offer comprehensive explanations, examples, and applications of game theory concepts, including coalitions, power dynamics, and decision-making in cooperative games. Dive into these texts to expand your understanding and explore fascinating real-world applications.

Trick: To assess the power dynamics within a coalition, consider conducting a power analysis by evaluating key players' resources, skill sets, and network influence. This analysis can provide valuable insights into the distribution of power and potential negotiation tactics.

Caveat: Power dynamics within coalitions can change over time and in response to external factors. It is essential to continuously reassess and adapt strategies to navigate evolving power dynamics in cooperative games.

Summary

In this section, we explored the concept of coalitions in cooperative games and the allocation of power within them. We discussed the formation of coalitions, the role of power in decision-making, and the impact of bargaining power on resource distribution. Real-life examples and challenges demonstrated the relevance and application of these concepts in various fields. By understanding coalitions and power dynamics, we can gain a deeper insight into the complex interactions that drive cooperative games.

The Prisoner's Dilemma Revisited: Cooperation vs. Defection

In this section, we will delve deeper into the classic game theory scenario known as the Prisoner's Dilemma. This game is often used to explore the tension between cooperation and defection in strategic decision-making. We will examine the core concepts, strategies, and real-world applications of this dilemma.

Understanding the Prisoner's Dilemma

The Prisoner's Dilemma is a model that captures the conflict between self-interest and mutual cooperation. It involves two individuals who are arrested for a crime and are held in separate cells. The prosecutor lacks conclusive evidence, so they offer each prisoner a deal: if one prisoner remains silent while the other confesses, the silent prisoner will receive a reduced sentence, while the confessing prisoner will be set free. However, if both prisoners confess, they will both receive moderate sentences. The dilemma lies in deciding whether to cooperate (remain silent) or defect (confess).

Strategy and Outcome Analysis

To analyze the outcome of the Prisoner's Dilemma, we use a payoff matrix. Let's assign numerical values to the different outcomes. Suppose the maximum gain is 5 units, the moderate sentence is 3 units, and the minimum gain (serving a long prison term) is 1 unit. The matrix would look as follows:

	Prisoner B Remain Silent	Confess
Prisoner A Remain Silent	(3, 3)	(5, 1)
Prisoner A Confess	(1, 5)	(2, 2)

In the payoff matrix, the first entry in each cell represents the payoff for Prisoner A, and the second entry represents the payoff for Prisoner B. The numbers correspond to the amount of utility or benefit they receive from each outcome.

The optimal strategy in the Prisoner's Dilemma depends on what the other player does. If both players cooperate (remain silent), they will both receive the moderate sentence, resulting in a total payoff of (3, 3). However, if one player defects (confesses) while the other cooperates, the defecting player will receive a higher payoff of (5, 1), also known as "sucker's payoff," and the cooperating player will receive the minimum payoff of (1, 5), known as "temptation payoff." If both players defect, they will both receive the "punishment payoff" of (2, 2).

Strategies in the Prisoner's Dilemma

In the classic version of the Prisoner's Dilemma, where the game is played once and players do not have the opportunity to communicate or engage in repeated interactions, the dominant strategy for each player is to defect. Defecting ensures that a player never receives the sucker's payoff and avoids the risk of receiving the minimum payoff.

However, if the game is played repeatedly or if there is a possibility of future interactions, strategies can change. Tit-for-tat is a popular strategy that promotes cooperation. It starts with cooperation and then mimics the opponent's previous move. If the opponent defects, the strategy switches to defection, manifesting a proportional response.

Real-World Applications

The Prisoner's Dilemma has numerous real-world applications across various fields. In economics, it helps explain market behavior, competitive pricing, and the dilemma faced by firms considering collusion or competition. In politics and international relations, it sheds light on arms races, negotiations, and conflicts where cooperation is essential.

An example of the Prisoner's Dilemma in action is the tragedy of the commons, where individuals pursuing their self-interest deplete shared resources, leading to the detriment of all. Another example is climate change, where nations must decide whether to cooperate by reducing emissions or defect by prioritizing their own economic growth.

Additionally, the Prisoner's Dilemma is used to study social dilemmas, such as public goods provision, prisoner's dilemma games in social networks, and the evolution of cooperation.

Ethical Considerations

The Prisoner's Dilemma raises ethical questions about cooperation and defection. It prompts us to examine our moral responsibilities in different situations, whether we prioritize self-interest or consider the collective good. In some cases, cooperating may be seen as an altruistic act, but one must also consider the risks associated with trusting others to cooperate.

Wrap Up

The Prisoner's Dilemma challenges the assumption that self-interest always leads to optimal outcomes. It highlights the tension between cooperation and defection and reveals the complexities of decision-making in strategic interactions. While defection may seem rational in a one-time game, repeated interactions and the potential for cooperation present alternative strategies. Understanding these dynamics is crucial for navigating negotiations, conflicts, and other strategic situations in various domains.

To delve deeper into game theory and its applications, continue reading this textbook. Prepare to question traditional assumptions, challenge your thinking, and enhance your decision-making skills through the lens of game theory.

Exercise: Consider a real-life scenario where the Prisoner's Dilemma can be applied. Describe the dilemma, the relevant players, and the potential outcomes. Discuss the strategies that each player might adopt and evaluate the ethical dimensions of the situation.

Finitely Repeated Games and Cooperation

In the previous section, we explored cooperative games and discussed how players can work together to achieve mutual benefits. However, in many real-world scenarios, cooperation is not a one-time deal. Instead, players are often engaged in repeated interactions where their decisions and actions can have long-term consequences.

In this section, we delve into finitely repeated games, which are a specific type of game where players engage in a fixed number of rounds of play. We will discuss the strategies and dynamics involved in these games, and how cooperation can be sustained even when faced with the temptation to defect.

The Prisoner's Dilemma Revisited

To understand finitely repeated games, it is essential to revisit the classic Prisoner's Dilemma. In this game, two suspects are arrested but lack sufficient evidence for a conviction. The prosecutor offers each suspect a deal: betray the other by confessing, and receive a reduced sentence if the other remains silent. If both confess, they each receive a moderate sentence. However, if both remain silent, they both receive a lighter sentence.

In a single round of the Prisoner's Dilemma, the dominant strategy is to betray the other, as it maximizes individual gain regardless of the opponent's decision. However, when the game is repeated, cooperation can become a viable option.

Cooperation Strategies in Finitely Repeated Games

One way to sustain cooperation in finitely repeated games is through the application of tit-for-tat strategy. Tit-for-tat starts with a cooperative move and then mimics the opponent's previous action in each subsequent round. It rewards cooperation with cooperation and punishes defection with defection.

Tit-for-tat has proven to be effective in a wide range of contexts. In the Prisoner's Dilemma, it creates a feedback loop of cooperation, as one player's betrayal in a previous round leads to a retaliatory response in the next round. Over time, this reciprocal behavior fosters cooperation despite the initial temptation to defect.

Another notable strategy is the grim trigger strategy, which starts with cooperation but switches to defection permanently if the opponent ever defects. This strategy provides a strong disincentive for defection as a single betrayal triggers an unrecoverable breakdown of cooperation.

Maintaining Cooperation: Strategies and Signals

While tit-for-tat and grim trigger are effective strategies, they rely on perfect information and a lack of mistakes. In reality, communication and signals play a crucial role in maintaining cooperation.

Signaling mechanisms allow players to convey their intentions and build trust. For instance, players can use costly signals to indicate their commitment to cooperation, such as investing time, resources, or reputation into a relationship. By signaling a willingness to cooperate, players can deter opportunistic behavior and foster long-term cooperation.

Moreover, reputation plays a vital role in repeated interactions. A player's reputation serves as a signal to others about past behavior and can influence future cooperation. A well-established reputation for cooperation can incentivize other players to reciprocate and encourages mutual trust.

Real-World Applications

Finitely repeated games have numerous real-world applications. One prominent example is the management of common-pool resources, such as fisheries or water reservoirs. In these scenarios, multiple stakeholders must decide how much to extract from a shared resource. By applying cooperation strategies and considering the long-term consequences of their actions, stakeholders can mitigate overexploitation and ensure the sustainable use of resources.

Another application is business negotiations and contracts. When companies engage in repeated interactions, such as in long-term supply-chain relationships, cooperation strategies can prove beneficial. By valuing long-term partnerships and employing cooperation mechanisms, companies can establish trust, reduce transaction costs, and foster mutually beneficial outcomes.

Exercise

Consider a business scenario where two companies are continuously bidding for contracts. They are engaged in a finitely repeated game where cooperation can lead to mutual benefits. Using the concepts discussed in this section, devise a cooperation strategy that enhances long-term cooperation and ensures better outcomes for both companies.

Hint: Consider strategies that involve signaling, reputation building, and the use of cooperation mechanisms like tit-for-tat.

Summary

Finitely repeated games provide a framework for studying cooperation in situations where players have repeated interactions. Strategies such as tit-for-tat and grim trigger have proven effective in sustaining cooperation over multiple rounds. Signaling, reputation, and trust-building mechanisms are crucial in maintaining cooperation in these games. Real-world applications of finitely repeated games include resource management and business negotiations. By understanding and applying the principles of finitely repeated games, we can navigate complex decision-making scenarios and achieve optimal outcomes through cooperation.

In the next section, we will explore the fascinating relationship between game theory and social networks. We will examine how the structure of social networks influences cooperation and strategic interactions. Join me as we unravel the interconnected nature of human interactions and the game theory principles that underpin them.

Repeated Games: Tit-for-Tat and Beyond

In the world of game theory, repeated games play a significant role in understanding strategic interactions. Unlike one-shot games where players make decisions based solely on immediate outcomes, repeated games allow players to consider the consequences of their actions in the long run. One classic strategy that has gained prominence in repeated games is the tit-for-tat strategy. In this section, we explore the concept of repeated games, delve into the details of the tit-for-tat strategy, and explore variations and extensions beyond tit-for-tat.

Understanding Repeated Games

Repeated games involve playing the same game multiple times in a sequence, allowing players to observe and learn from each other's actions. The decisions made in each round can depend on the outcomes of previous rounds, creating a strategic dynamic between players. The number of repetitions can be predetermined or indefinite, adding an element of uncertainty and strategic thinking.

Repeated games can help uncover cooperation and establish trust between players. They also provide a framework for analyzing strategies that balance immediate gains with long-term benefits.

The Tit-for-Tat Strategy

The tit-for-tat strategy is one of the most well-known and successful strategies in repeated games. It gained recognition through its performance in Robert Axelrod's famous Prisoner's Dilemma tournaments.

The basic idea behind tit-for-tat is simple: start by cooperating and then mimic your opponent's previous move in subsequent rounds. This strategy encourages cooperation and reciprocates both good and bad actions, fostering trust and promoting cooperation.

Tit-for-tat has several advantages. First, it is easy to understand and implement, requiring minimal cognitive effort. Second, it is forgiving, allowing players to recover from occasional mistakes without holding grudges. Finally, it is transparent and straightforward, making it easy for opponents to recognize and adapt to the strategy.

Variations and Extensions

While tit-for-tat has proven to be a successful strategy in many situations, variations and extensions have been developed to address specific challenges and to exploit potential weaknesses.

One variation is tit-for-two-tats, where a player only retaliates after their opponent defects twice in a row. This strategy offers a more forgiving approach, allowing for occasional errors while maintaining the importance of cooperation.

Another variation is forgiving tit-for-tat, where a player forgives their opponent after a single defection. This strategy promotes cooperation by giving a second chance to defectors, fostering reconciliation and long-term collaboration.

In addition to variations, other strategies have emerged that go beyond simple retaliation. For example, grim trigger strategy involves initially cooperating but permanently defecting if the opponent defects at any point. This strategy aims to deter opponents from betraying cooperation.

Real-World Applications

Repeated games and the tit-for-tat strategy have real-world applications in a wide range of fields. In economics, repeated interactions between firms in an oligopoly can be modeled using repeated games, allowing for strategic analysis of pricing and competition.

In international relations, repeated games provide a framework for understanding alliances, negotiations, and conflicts between countries. The tit-for-tat strategy can be applied to build trust and promote cooperation in diplomatic engagements.

Furthermore, the tit-for-tat strategy has found practical applications in computer science, particularly in designing algorithms for network protocols, cybersecurity, and autonomous systems. By establishing a reputation-based mechanism, systems can incentivize cooperation and deter malicious behavior.

Exercise

Consider a repeated game scenario where two players, Alice and Bob, can choose to cooperate (C) or defect (D) in each round. Their payoffs are as follows:

Alice\Bob	C	D
C	4, 4	0, 5
D	5, 0	1, 1

Answer the following questions:
1. What would be the outcome if both players follow the tit-for-tat strategy? 2. How might the outcome change if Alice and Bob adopt the grim trigger strategy? 3. Can you think of any real-world scenarios where the tit-for-tat strategy would be effective? Explain your reasoning.

Conclusion

Repeated games provide a deeper understanding of strategic interactions by considering the consequences of actions over time. The tit-for-tat strategy, with its simplicity and effectiveness, has become a widely recognized approach to promoting cooperation. Variations and extensions of this strategy offer flexibility in different scenarios.

By analyzing repeated games and understanding strategies like tit-for-tat, we gain valuable insight into fields ranging from economics and international relations to computer science and beyond. The applications of these concepts are diverse, offering practical solutions to common challenges in real-world interactions.

Applications of Cooperative Game Theory in Business and International Relations

Cooperative game theory provides valuable insights into strategic decision-making, resource allocation, and negotiation in both business and international relations. It helps us understand how individuals and organizations can mutually benefit from cooperation, overcome conflicts, and achieve optimal outcomes. In this section, we will explore various real-world applications of cooperative game theory in these fields.

Strategic Alliances in Business

Strategic alliances are agreements between two or more firms that join forces to achieve mutual goals and enhance their competitive advantage.

Cooperative game theory offers a powerful framework to analyze these partnerships and guide decision-making.

One common application is in joint ventures, where companies combine resources, expertise, and technologies to enter new markets or develop innovative products. Game theory helps firms assess the risks and rewards of collaboration, identify optimal strategies, and define the distribution of benefits. For example, consider a situation where two pharmaceutical companies collaborate to develop a new drug. They must negotiate the financing, intellectual property rights, and revenue sharing to ensure a fair and beneficial partnership.

Cooperative game theory also aids in understanding supply chain management. Companies within a supply chain must cooperate to optimize efficiency, reduce costs, and improve customer satisfaction. Game theory models help identify stable coalitions and enable fair allocation of costs and benefits among the players. This ensures that all parties involved have an incentive to cooperate rather than act competitively.

Negotiations and Conflict Resolution

Cooperative game theory plays a crucial role in negotiations and conflict resolution by providing a structured approach to problem-solving and decision-making. Whether it is a business deal or a peace treaty between nations, cooperative strategies are essential for creating sustainable agreements.

In negotiations, understanding the potential gains from cooperation and the trade-offs involved is crucial. Game theory helps in analyzing the bargaining power and leverage of each party, identifying possible outcomes, and designing strategies that lead to mutually beneficial agreements.

For instance, consider a negotiation between a labor union and a company regarding wage increases and working conditions. Cooperative game theory allows both sides to assess their respective interests, explore possible compromises, and find a solution that maximizes overall satisfaction.

In international relations, cooperative game theory is used to analyze various diplomatic scenarios such as arms control agreements, trade negotiations, and climate change agreements. By studying the incentives and potential gains from cooperation, policymakers can design strategies to encourage collaboration and resolve conflicts effectively.

Cooperative Game Theory in Business Ethics

Cooperative game theory also offers insights into ethical decision-making in business. It provides a framework to analyze ethical dilemmas, guide moral reasoning, and promote fairness and trust among stakeholders.

For example, consider a situation where a pharmaceutical company has discovered a life-saving drug. By analyzing the potential gains from cooperation and the ethical implications of various pricing strategies, game theory can help the company make an informed decision. It allows them to balance the need for profitability with considerations of accessibility and fairness, ensuring that the drug benefits as many people as possible.

Cooperative game theory can also be applied to corporate social responsibility initiatives. By modeling the interactions between businesses, consumers, and society, game theory helps identify cooperative strategies that align the interests of stakeholders and promote sustainable development.

Diplomacy and Global Cooperation

In international relations, cooperative game theory plays a critical role in understanding and facilitating global cooperation. It provides a framework for analyzing complex issues such as arms races, climate change, and humanitarian crises.

Game theory models can help policymakers identify stable coalitions, assess the benefits of cooperation, and design strategies to address common challenges. For example, in climate change negotiations, game theory helps countries understand the incentives for emitting less carbon dioxide and reach agreements that promote sustainable development.

Cooperative game theory also aids in understanding alliances and collective security. By analyzing the potential gains from cooperation and the risks of defection, policymakers can design strategies to foster trust, maintain stability, and prevent conflicts.

Limitations and Caveats

While cooperative game theory offers valuable insights, it is essential to recognize its limitations and caveats. One limitation is the assumption of rationality, which may not always hold in real-world situations. In practice, individuals and organizations may exhibit bounded rationality or be influenced by emotions and biases, impacting their strategic decisions.

Additionally, achieving cooperation in games can be challenging due to the potential for defection or the presence of multiple equilibria. Building trust among players and creating credible commitments are critical for successful cooperation.

It is also important to consider the contextual factors that may influence cooperative behavior. Cultural differences, power dynamics, legal frameworks, and institutional arrangements can significantly impact the success of cooperative strategies in different settings.

Conclusion

Cooperative game theory provides valuable tools for analyzing strategic decision-making, resource allocation, and negotiation in business and international relations. By understanding the principles of cooperation, stakeholders can identify mutually beneficial strategies, resolve conflicts, and promote sustainable development.

Through strategic alliances, negotiations, and ethical considerations, businesses can leverage cooperative game theory to enhance their competitive advantage, improve supply chain management, and foster stakeholder trust. In international relations, cooperative game theory aids in diplomacy, global cooperation, and conflict resolution, contributing to a more stable and prosperous world.

By applying cooperative game theory, individuals and organizations can play smart and achieve optimal outcomes by cooperating rather than acting solely in their self-interest.

Game Theory in Social Networks

Social Network Analysis and Game Theory

Network Structure and Behavior

In the realm of game theory, the study of network structure and behavior plays a crucial role in understanding how individuals interact and make decisions in social networks. A social network can be defined as a set of actors, or nodes, connected by relationships, or edges. These relationships can represent various forms of social ties, such as friendship, collaboration, or information flow.

Network structure refers to the patterns of connections within a network. It provides insights into the underlying social dynamics and influences how information, behaviors, and resources spread within the network. Understanding network structure is vital for analyzing strategic interactions and predicting outcomes in diverse scenarios, ranging from the diffusion of innovations to the spread of rumors or diseases.

Centrality and Influence in Social Networks

One of the fundamental concepts in network analysis is centrality, which measures the relevance or importance of a node within a network. Centrality captures the idea that not all nodes have equal influence or significance. Several measures of centrality exist, each focusing on different aspects of importance.

One widely used centrality measure is degree centrality, which quantifies the number of connections a node has. In social networks, a node with a high degree centrality is often referred to as a "hub" or a "connector," as it has many direct connections to other nodes. Hubs play a crucial role in information diffusion, as they have the potential to reach a large number of individuals quickly.

Another measure of centrality is betweenness centrality, which considers the extent to which a node lies on the shortest paths between other pairs of nodes. Nodes with high betweenness centrality act as bridges, facilitating the flow of information or resources between different parts of the network. Removing these nodes can have a significant impact on the network's connectivity and communication efficiency.

Additionally, closeness centrality measures how close a node is to all other nodes in the network. Nodes with high closeness centrality can quickly access information from the rest of the network. They are often well-positioned to disseminate information efficiently and exert influence over others. Closeness centrality is particularly relevant in contexts where speed of information transmission is critical.

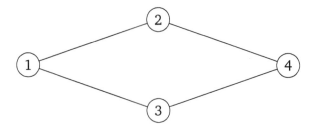

Figure 0.1: Example social network.

To illustrate the concept of centrality, consider the example social network shown in Figure 0.1. Node 1 has the highest degree centrality since it is directly connected to two other nodes. However, nodes 2 and 3 have higher betweenness centrality because they must be traversed to reach node 4 from node 1. Node 4, on the other hand, has the highest closeness centrality due to its direct connections with nodes 2 and 3.

Small World Phenomenon and Social Connectivity

The small world phenomenon is a fascinating aspect of network structure that highlights the interconnectedness of individuals within a network. It posits that, on average, any two individuals in a social network are only a few steps away from each other, despite the vast size of the network.

One classic experiment demonstrating the small world phenomenon is Stanley Milgram's "Six Degrees of Separation" study. Milgram asked participants to send letters to a target person, but they could only forward the letter to someone they knew personally. The study found that, on average, it took only six steps for the letter to reach the target person. This experiment showcased the power of social connections and the unexpected short paths that exist between individuals.

The small world phenomenon has significant implications for the spread of information, ideas, and even behaviors. It suggests that individuals can quickly disseminate information or make connections through their networks, enabling rapid diffusion and influencing collective behavior.

Homophily and Social Influence

In social networks, individuals tend to form connections with others who are similar to themselves, a phenomenon known as homophily. Homophily is driven by shared interests, values, attitudes, or demographic characteristics. For example, people often befriend others of similar age, occupation, hobbies, or political beliefs.

Homophily plays a crucial role in shaping the dynamics of social influence within networks. When individuals are connected to others who share their preferences or opinions, they are more likely to be influenced by their peers' behaviors and choices. This can lead to the formation of echo chambers, where individuals reinforce and amplify each other's beliefs, potentially resulting in polarization within a network.

Understanding homophily is essential for analyzing the spread of information, attitudes, and behaviors in social networks. By accounting for social influence and the effects of similarity, researchers can develop strategies to promote positive behaviors or mitigate the impact of misinformation.

Network Dynamics and Game Theory

The structure of a social network is not fixed but constantly evolving due to various dynamics, such as the formation of new connections, the dissolution of existing ties, or changes in individuals' attributes. Game theory provides a powerful framework to study how individual decisions and behaviors shape the structure and dynamics of a network.

Erdős-Rényi random graphs and Barabási-Albert preferential attachment models are two common mathematical models used to simulate network dynamics. They allow researchers to investigate how different rules and strategies for forming connections can lead to specific network properties, such as the emergence of hubs or the presence of tightly knit clusters.

Game theory also helps analyze strategic interactions in network formation. Individuals face decisions about whether to connect, sever ties, or choose specific partners. These decisions are often influenced by the anticipated benefits, costs, or risks associated with particular connections. By considering the strategic behavior of individuals, game theory enriches our understanding of network formation processes and provides insights into the emergence of collective phenomena.

Diffusion and Adoption in Social Networks

One of the fundamental processes taking place in social networks is the diffusion of information, innovations, or behaviors. Diffusion refers to the spread of something through a network, where individuals adopt a new idea or behavior based on their connections to others who have already adopted it.

Understanding diffusion processes is crucial for various domains, such as public health, marketing, and social policy. Game theory can shed light on the dynamics of diffusion by considering how individuals' decisions to adopt a new behavior interact and influence others in the network.

A classic model for studying diffusion is the threshold model. In this model, individuals have a threshold that determines when they adopt a behavior. If the number of their neighbors who have already adopted the behavior exceeds their threshold, they are likely to adopt it as well. This model provides insights into the critical mass needed for widespread adoption and the role of network structure in accelerating or hindering diffusion.

Cooperation and Trust in Social Networks

Cooperation is an essential aspect of many social interactions. In social networks, individuals often face collective dilemmas, where the pursuit of individual interests conflicts with the overall group's welfare. Game theory offers valuable tools and concepts to understand cooperation and trust in social networks.

One well-known game used to study cooperation is the Prisoner's Dilemma. It captures the tension between individual rationality and collective well-being. In the context of social networks, multiple versions of the Prisoner's Dilemma can be explored, taking into account how individual decisions to cooperate or defect affect the entire network.

Network reciprocity is a mechanism that can promote cooperation in social networks. It posits that individuals are more likely to cooperate with their neighbors, as their future interactions are more probable. By examining different strategies, such as tit-for-tat or generous tit-for-tat, researchers can gain insights into how cooperation can be sustained even in the presence of selfish individuals.

Trust is another crucial element in social networks, as it enables individuals to rely on each other's actions and make informed decisions. Game theory provides insights into trust-building mechanisms and the conditions under which trust can emerge and be maintained in social networks. Trust plays a fundamental role in various contexts, from online communities to business partnerships and international relations.

Game Theoretic Models of Opinion Formation

Opinions and attitudes can spread rapidly through social networks, influencing individuals' beliefs and behaviors. Game theoretic models have been developed to study opinion formation and its dynamics within social networks.

Opinion dynamics models often combine game theory with elements of social influence, where individuals update their opinions based on the opinions of their neighbors. These models allow researchers to investigate how initial conditions, network structure, and individual decision-making rules shape the collective opinion that emerges.

One notable model is the DeGroot model, in which individuals iteratively update their opinions by averaging the opinions of their neighbors. The DeGroot model offers insights into long-term convergence or polarization of opinions based on the network's structure and initial diversity of opinions.

Real-world applications of opinion formation models include predicting the adoption of innovations, understanding the dynamics of collective decision-making, and analyzing the impact of social media on public opinion.

Real-World Applications of Game Theory in Social Networks

The study of network structure and behavior, combined with game theory, has extensive real-world applications across various domains. Some notable examples include:

- Public Health: Understanding how diseases spread through social networks helps in developing effective containment strategies and vaccination campaigns.

- Marketing and Advertising: Analyzing the structure of social networks assists in identifying key influencers and designing targeted advertising campaigns for maximum impact.

- Online Social Platforms: Social media platforms leverage network structure and dynamics to personalize content recommendations, connect users with similar interests, and detect malicious behavior.

- Collaborative Innovation: Analyzing social networks within organizations aids in fostering knowledge sharing, collaboration, and innovation.

By applying game theory concepts and network analysis, researchers and practitioners can gain valuable insights into the complex dynamics of social networks and develop strategies for effective decision-making, policy design, and behavior change campaigns.

Summary

In this section, we explored the world of social networks and their relevance in game theory. We discussed the importance of network structure and the concept of centrality in understanding network dynamics and influence. Furthermore, we explored the small world phenomenon, homophily, and different dynamics that shape social connectivity.

We also delved into the role of game theory in analyzing network formation, diffusion processes, cooperation, trust, and opinion formation within social networks. Lastly, we highlighted various real-world applications where game theory and network analysis have proven to be powerful tools.

As we move forward in this book, it is essential to grasp the interconnectedness of individuals within social networks and the strategic decision-making that shapes their behavior. By understanding the structure and behavior of social networks, we can navigate the complexities of real-world interactions and make smarter decisions. Remember, it's not just about playing hard but playing smart!

Centrality and Influence in Social Networks

In social networks, understanding the dynamics of influence and centrality is crucial to unravel the different ways people connect with each other, share information, and shape their opinions. This section explores the concepts of centrality and influence in social networks, highlighting their significance and providing methods to measure and analyze them.

Understanding Centrality

Centrality refers to the importance and prominence of an individual or node within a social network. It provides insights into the structural position and influence of a person within their social environment. There are several measures of centrality, each capturing different aspects of a node's centrality. Let's explore some commonly used centrality measures:

- **Degree Centrality:** This measure quantifies the number of connections a node has in a network. Nodes with higher degree centrality are considered more central, as they have a wider reach and can access a greater number of other nodes.

- **Closeness Centrality:** Closeness centrality assesses how close a node is to all other nodes in terms of geodesic distance. It measures the speed at which information can spread from a node to others in the network. Nodes with higher closeness centrality have more direct access to other nodes and can disseminate information quickly.

- **Betweenness Centrality:** Betweenness centrality identifies nodes that act as intermediaries or bridges between different parts of a network. Nodes with higher betweenness centrality have the potential to control information flow and act as gatekeepers. They are critical for maintaining connectivity within a network.

- **Eigenvector Centrality:** Eigenvector centrality evaluates the influence of a node based on the centrality of its connected neighbors. Nodes with higher eigenvector centrality are well-connected to other influential nodes, making them more important in the network. This measure captures the idea of "being influential by association."

Understanding centrality measures allows us to identify key players, influential voices, and influential nodes in social networks. However, it is essential to note that centrality does not always equate to influence, as influence can be more complex and multidimensional.

Measuring Influence

Measuring influence in social networks is a challenging task, as influence can manifest in different forms and contexts. Nevertheless, researchers have developed various methods to capture and quantify influence. Let's discuss a few widely used approaches:

- **Information Diffusion Models:** By analyzing the spread of information or behaviors within a network, we can estimate the influence of different nodes. Models such as the Independent Cascade Model and the Linear Threshold Model simulate the diffusion process and identify influential nodes based on their ability to initiate cascades of information.

- **PageRank Algorithm:** Developed for ranking web pages, the PageRank algorithm can also be applied to social networks. It evaluates the importance of a node based on the volume and quality of incoming links. Nodes with higher PageRank scores are considered more influential.

- **Opinion Leaders:** Opinion leaders are individuals who are recognized and respected for their expertise or knowledge in a particular area. Identifying opinion leaders within a social network can provide insights into influential individuals.

- **Social Network Analysis:** Analyzing the structure and dynamics of social networks can reveal key players and communities. Key players often have a significant influence on the opinions and behaviors of others. Identifying these actors can help understand influence dynamics.

Measuring influence is a complex task, and it often requires a combination of quantitative and qualitative methods. Additionally, influence can vary across different domains and contexts, making it essential to adapt measurement techniques accordingly.

Applications of Centrality and Influence Analysis

Understanding centrality and influence in social networks has a wide range of applications. Let's explore some areas where these concepts play a vital role:

- **Marketing and Advertising:** Identifying influential individuals or opinion leaders in social networks can help target advertising campaigns more effectively. These individuals can act as brand ambassadors and encourage others to adopt certain products or services.

- **Public Health:** Studying the spread of diseases or the adoption of healthy behaviors in social networks can help design targeted interventions that leverage influential individuals to promote positive health outcomes.

- **Political Campaigns:** Analyzing influence dynamics in social networks can aid political campaigns in identifying key endorsers or influencers who can mobilize supporters and shape voter opinions.

- **Information Dissemination:** Understanding the flow of information in social networks can facilitate the effective dissemination of news and facts. Identifying central nodes can help ensure accurate information reaches a wide audience.

- **Social Movements:** Social movements often rely on influential nodes within social networks to spread messages and mobilize participants. Identifying these key players can help understand the dynamics of social change.

It is important to note that centrality and influence analysis should be used ethically and responsibly. Privacy considerations, potential biases, and unintended consequences should always be taken into account when applying these methods in real-world scenarios.

Summary

Centrality and influence are fundamental concepts in social network analysis. Centrality measures help identify important nodes within a network, while influence analysis aims to understand the dynamics of information spread and behavior adoption. These concepts find applications in a variety of domains, including marketing, public health, politics, and social movements. As social networks continue to shape our interactions and shape societal dynamics, understanding and leveraging centrality and influence will only grow in importance.

Exercises

1. Reflect on your own social network and identify individuals who exhibit different types of centrality (e.g., degree centrality, closeness centrality, betweenness centrality). Discuss the potential influence of these individuals.

2. Choose a social network platform (e.g., Facebook, Twitter, Instagram) and examine how they employ algorithms to measure and display influence metrics. Critically analyze the limitations and biases associated with these algorithms.

3. Research a public health campaign that leveraged influential individuals or opinion leaders within social networks. Discuss the effectiveness of these strategies and any challenges faced.

4. Explore a real-world case where a social movement relied on influential nodes within a social network to mobilize participants. Analyze the strategies employed and the role of influence in the success of the movement.

5. Design a hypothetical marketing campaign that utilizes influential individuals within a social network to promote a new product. Outline the key steps and considerations in identifying and engaging these influencers.

Further Reading

- Borgatti, S.P., Everett, M.G., & Johnson, J.C. (2018). *Analyzing Social Networks*. SAGE Publications.

- Watts, D.J. (2004). *Six Degrees: The Science of a Connected Age*. W. W. Norton & Company.

- Barabási, A.L. (2002). *Linked: The New Science of Networks*. Basic Books.

- Newman, M.E.J. (2018). *Networks*. Oxford University Press.

Conclusion

Centrality and influence are crucial concepts in understanding the dynamics of social networks. By exploring different centrality measures and influence analysis techniques, we can gain insights into the structure, behavior, and impact of social networks. Applying these concepts to various domains allows us to make informed decisions and design targeted interventions. As social networking continues to evolve, so too must our understanding of centrality and influence in unraveling the intricacies of human connections.

Small World Phenomenon and Social Connectivity

Introduction: In the vast landscape of social networks, have you ever wondered why it feels like everyone is connected somehow? Well, that's because of a fascinating concept called the small world phenomenon. It reveals how surprisingly small and interconnected our social connections can be. In this section, we will explore the small world phenomenon and its implications for social connectivity.

Understanding the Small World Phenomenon: The small world phenomenon, popularized by Stanley Milgram's famous "six degrees of separation" experiment, refers to the idea that any two people in the world can be connected through a short chain of social connections. It signifies that social networks have a remarkable tendency for short distances between individuals, regardless of their geographical or cultural differences.

Network Structure and Behavior: To understand the small world phenomenon, we need to delve into the structure and behavior of social networks. Social networks consist of nodes (individuals) and edges (connections between individuals). The arrangement of these connections gives rise to the network structure, which can have a significant impact on the dynamics of information flow, influence, and cooperation within the network.

Centrality and Influence in Social Networks: Within social networks, certain individuals hold a key position in the network structure known as centrality. Centrality measures the importance or influence of a node based on its connectivity and position in the network. Nodes with high centrality, such as those with many connections or bridging different communities, have the power to spread information or influence opinions more effectively.

Small World Networks: The small world phenomenon is often explained through the concept of small world networks. These networks possess a combination of high local clustering (nodes connected to nearby neighbors) and short average path lengths (low number of steps required to connect any two nodes). Small world networks are characterized by the presence of so-called "hubs" - highly connected nodes that facilitate efficient communication and information flow throughout the network.

The "Six Degrees of Separation" Game: To gain a better understanding of the small world phenomenon, let's play a game.

Imagine you have to find a connection between yourself and a celebrity, say Elon Musk. You can reach out to your friends, who can then connect you with their friends, and so on. The challenge is to find the shortest chain of connections that links you to Elon Musk. This game demonstrates how the small world phenomenon operates in practice, as the vast majority of participants are able to establish a connection within six degrees of separation.

Homophily and Social Influence: Another important aspect of social connectivity is homophily - the tendency of individuals to associate with others who share similar characteristics, such as interests, beliefs, or backgrounds. Homophily plays a role in shaping social networks, as individuals are more likely to form connections with others who are like themselves. This phenomenon can have both positive aspects, fostering a sense of community, and negative aspects, reinforcing echo chambers and limiting exposure to diverse opinions.

Network Dynamics and Formation: Social networks are not static entities but constantly evolve over time through various mechanisms. These dynamics can include the formation and dissolution of connections, the emergence of network communities, and the influence of external factors. By studying network dynamics, researchers gain insights into how social networks shift and adapt, leading to changes in connectivity patterns and the overall structure of the network.

Real-World Applications: The small world phenomenon has real-world applications in various fields, including sociology, marketing, and public health. By understanding the structure and dynamics of social networks, organizations can efficiently disseminate information, target specific groups, and even predict the spread of diseases. For example, in a viral marketing campaign, identifying influential individuals or hubs within a social network can greatly amplify the impact of advertising efforts.

Conclusion: The small world phenomenon reveals the incredible interconnectedness of our social networks, demonstrating that we are all much closer than we may think. By studying social connectivity and the small world phenomenon, we gain insights into how information, influence, and cooperation spread through networks. This knowledge enables us to navigate and leverage the power of social networks in various domains, ultimately shaping our interactions and understanding of the world around us.

Further Reading: - "Connected: The Surprising Power of Our Social Networks and How They Shape Our Lives" by Nicholas A. Christakis and James H. Fowler - "Networks, Crowds, and Markets: Reasoning About a Highly Connected World" by David Easley and Jon Kleinberg - "Six Degrees of Separation: The Science of a Connected Age" by Duncan J. Watts. - "Linked: The New Science of Networks" by Albert-László Barabási.

Practice Exercises: 1. Reflect on your own social network and identify any hubs or highly connected individuals. How do these individuals influence the flow of information within the network? 2. Conduct a small survey among your friends to investigate if there are any common interests or characteristics that led to the formation of connections in your network. Discuss the findings and implications. 3. Research a real-world application of the small world phenomenon in marketing, public health, or another field of interest. Write a short report outlining the insights and benefits gained from understanding social connectivity.

Homophily and Social Influence

In social networks, individuals often form connections with others who are similar to them. This phenomenon, known as homophily, plays a crucial role in shaping social interactions. Homophily can be observed in various aspects of our lives, from friendships and romantic relationships to shared interests and beliefs. Understanding homophily is essential in studying social influence and the dynamics of social networks.

Defining Homophily

Homophily refers to the tendency of individuals to associate and form relationships with others who are similar to them in certain attributes or characteristics. These similarities can be based on various factors such as demographics (e.g., age, gender, race), socioeconomic status, interests, attitudes, and beliefs. For instance, individuals with similar political views are more likely to form connections and engage in political discussions with each other.

Mechanisms of Homophily

Homophily can arise through different mechanisms, including selection, socialization, and influence.

Selection: In some cases, individuals actively seek out and choose to connect with others who are similar to them. This can be driven by shared interests, values, or personal preferences. For example, someone who enjoys playing video games may actively seek out online communities or gaming groups to connect with like-minded individuals.

Socialization: Homophily can also occur through socialization processes, where individuals adopt the characteristics and behaviors of those they interact with. Through socialization, individuals may become more similar to their social contacts over time. For example, a person who joins a sports team may adopt the habits and interests of their teammates.

Influence: Homophily can be a result of social influence, where individuals align their beliefs, attitudes, and behaviors with those of their social contacts. This can be driven by a desire for social approval or a need to conform to group norms. For example, someone may change their fashion style to match that of their friends, or start listening to a new music genre because their peers enjoy it.

Social Influence in Social Networks

Social influence refers to the process by which individuals' thoughts, feelings, and behaviors are affected by the actions and opinions of others. It is a fundamental aspect of human interactions and plays a significant role in shaping individual and group behaviors within social networks.

Informational Influence: One form of social influence is informational influence, where individuals conform to the opinions or behaviors of others because they believe those others have valuable information. When individuals are uncertain about the correct course of action, they may look to their social contacts for guidance. This is particularly true in situations where the consequences of a decision are ambiguous or unfamiliar.

Normative Influence: Normative influence occurs when individuals conform to the expectations and norms of a social group or society. People have an innate desire to be accepted and approved by others, which often leads them to conform to social norms. Normative influence can be a powerful force in shaping behaviors, attitudes, and opinions. For example, individuals may adopt certain political views or religious beliefs to align with those of their social group.

Homophily and Social Influence in Action: Echo Chambers

Homophily and social influence can have both positive and negative consequences within social networks. One notable phenomenon that arises due to these mechanisms is the formation of echo chambers.

An echo chamber refers to a situation where individuals are exposed only to information and opinions that reinforce their existing beliefs and values. This occurs when people surround themselves with like-minded individuals and engage in selective exposure to information that aligns with their preexisting views. Echo chambers can lead to a reinforcement of existing attitudes, polarization of opinions, and a reduced willingness to consider alternative perspectives.

In online platforms and social media, echo chambers can become particularly prominent due to algorithms that tailor content based on users' preferences and past behaviors. These algorithms aim to maximize user engagement by recommending content that is more likely to elicit positive responses. As a result, individuals may be continuously exposed to information and viewpoints that validate their beliefs, further reinforcing homophily and creating information bubbles.

Breaking out of echo chambers and promoting diverse perspectives becomes challenging when there is limited exposure to alternative viewpoints. This can lead to a lack of critical thinking, reduced open-mindedness, and an increase in misinformation or confirmation bias. Addressing echo chambers requires conscious efforts to seek out diverse sources of information, engage in respectful discussions with those holding different views, and promoting media literacy.

Real-World Applications

Understanding homophily and social influence has practical applications in various fields, including marketing, public health, and politics.

Marketing: Marketers often leverage the principles of homophily and social influence to target specific consumer segments. By understanding the preferences and characteristics of their target audience, marketers can design campaigns that resonate with their customers and use social proof to drive sales. Influencer marketing, for example, capitalizes on the power of social influence by leveraging individuals with large followings to promote products or services.

Public Health: In public health campaigns, knowledge of homophily and social influence can inform strategies for behavior change. By identifying influential individuals within a community, interventions can be designed to leverage their social networks and promote positive health behaviors. This approach has been successful in areas such as HIV prevention, where peer educators use their influence to disseminate accurate information and encourage safer practices.

Politics: Homophily and social influence are crucial in political campaigns and elections. Voters are often influenced by the opinions and behaviors of their social contacts, making peer-to-peer communication an effective strategy. Understanding the dynamics of homophily and social influence can help campaign strategists identify key influencers and target messages that resonate with specific voter groups.

Conclusion

Homophily and social influence are fundamental aspects of social networks. Through homophily, individuals form connections with others who are similar to them, and through social influence, they are influenced by the opinions, attitudes, and behaviors of their social contacts. Understanding these mechanisms is vital in studying social dynamics, echo chambers, and the diffusion of information within social networks. By recognizing the power of homophily and its influence on individual and group behaviors, we can better navigate the complexities of social interactions in various domains.

Network Dynamics and Game Theory

In this section, we explore the fascinating intersection between network dynamics and game theory. Network dynamics refers to the study of how complex systems evolve over time through interactions between interconnected entities. Game theory, on the other hand, provides a framework for analyzing strategic decision-making in various competitive and cooperative situations. By combining these two fields, we can gain insights into how network structures impact the behavior and outcomes of individuals within those networks.

Understanding Network Dynamics

Network dynamics examines the interactions and relationships between nodes (entities) in a network and how these connections evolve over time. A network can represent social relationships, technological systems, biological processes, or any other system in which entities are connected in some way. The behavior and characteristics of these entities depend not only on their individual attributes but also on the structure of the network itself.

Implications of Network Structure

The structure of a network plays a crucial role in determining its dynamics. Different network topologies can exhibit varying levels of connectivity, centrality, and clustering. For example, a highly connected network with many interconnections allows for information and influence to spread quickly. In contrast, a sparse network with limited connections may impede the flow of information.

Centrality refers to the relative importance of a node within a network. Nodes with high centrality have a significant influence on the overall dynamics of the network. They act as hubs that facilitate the transfer of information, resources, or influence between various nodes. Understanding centrality is crucial in identifying key players and predicting the behavior of the network as a whole.

Furthermore, clustering, which refers to the tendency of nodes to form tightly connected clusters or communities, can have a significant impact on network dynamics. In a social network, for instance, individuals tend to associate more closely with others who share similar interests or characteristics. These clusters can affect the diffusion of information, the spread of behaviors, and the formation of social norms within the network.

Game-Theoretic Approach to Network Dynamics

Game theory provides a powerful tool for analyzing strategic interactions within a network. By modeling the interactions between nodes as games, we can gain insights into how individual behavior and network structure influence each other. The strategic decisions made by individual nodes can have ripple effects throughout the network, affecting the behaviors and outcomes of other nodes.

In game theory, each node is considered a player that aims to maximize its own utility based on the actions of other players. The interactions between nodes can be either competitive (zero-sum) or cooperative (non-zero-sum). Players must strategically assess their choices, taking into account the potential responses of other players and the overall structure of the network.

Examples of Network Dynamics and Game Theory

One example of the application of network dynamics and game theory is the study of viral marketing campaigns on social media. Companies and advertisers seek to maximize the spread of information or influence within a network of users. By understanding the structure of social networks and the incentives that motivate individuals to share content, game theory can guide the design of effective strategies to encourage viral behavior.

Another example comes from the field of epidemiology. The spread of contagious diseases can be modeled as a game played between individuals in a network. By analyzing the incentives and behaviors of individuals, researchers can gain insights into the dynamics of disease transmission and the effectiveness of various intervention strategies, such as vaccination campaigns or social distancing measures.

Challenges and Future Directions

Studying network dynamics and game theory poses several challenges. The complexity and size of real-world networks can be daunting, requiring the development of scalable computational methods. Furthermore, accurately capturing individual behaviors and preferences within a game-theoretic framework presents a constantly evolving challenge.

The future of network dynamics and game theory lies in the integration of empirical data and computational modeling approaches. The availability of large-scale datasets and advances in network analysis techniques allow for a more nuanced understanding of how network structure influences behavior. Combining these insights with game-theoretic models can provide a powerful framework for predicting and shaping the dynamics of complex systems.

Summary

In this section, we delved into the fascinating topic of network dynamics and its relationship with game theory. We explored how the structure of networks impacts the behavior of individuals within them and how game theory provides a framework for analyzing strategic decision-making in networked settings. We discussed the implications of network structure, the game-theoretic approach to network dynamics, and provided examples of real-world applications. Finally, we highlighted the challenges and future directions in this exciting field. The combination of network dynamics and game theory offers a powerful toolkit for understanding and shaping the complex dynamics of interconnected systems.

Network Equilibrium and Strategy Evolution

In the previous section, we discussed the dynamics of social networks and how game theory can be applied to understand the behavior of individuals within these networks. Now, let's delve deeper into the concept of network equilibrium and how strategies evolve within social networks.

Understanding Network Equilibrium

Network equilibrium refers to a state where the behavior of individuals and the structure of the network are mutually reinforcing. In simpler terms, it's a point where the strategies adopted by individuals are in balance with the social connections they have within the network.

To understand network equilibrium, we need to consider two important factors: strategy and centrality. Strategy refers to the behavior or decision-making approach that individuals adopt within the network, while centrality represents the extent of influence an individual has over others in the network.

In a network equilibrium, individuals strategically adapt their behavior based on the behavior of their neighbors and their own centrality within the network. This adaptation process eventually stabilizes the network structure and the strategies employed by individuals.

Evolutionary Dynamics and Strategy Evolution

Strategy evolution in social networks occurs through an iterative process where individuals learn from and imitate the strategies of others. This process is influenced by two key mechanisms: social learning and evolutionary dynamics.

Social learning occurs when individuals observe the behavior of their neighbors and adopt strategies that seem successful or advantageous. This imitation-based learning process can lead to the emergence of popular strategies within the network.

On the other hand, evolutionary dynamics are driven by individual fitness and performance. Individuals with successful strategies tend to have better fitness, meaning they are more likely to pass on their strategies to future generations. Over time, this can lead to the dominance of certain strategies within the network.

The interplay between social learning and evolutionary dynamics shapes the evolution of strategies within social networks. As individuals imitate successful strategies and those strategies gain prominence, they become more prevalent in the network, creating a feedback loop that reinforces their adoption.

Network Equilibrium and Strategy Diversity

While network equilibrium suggests a stable state, it's important to note that diversity of strategies can coexist within a network. In fact, having a diverse range of strategies can be advantageous for the overall resilience and adaptability of the network.

In a network with high strategy diversity, individuals have access to a wider range of information and decision-making approaches. This diversity allows for more exploration and innovation within the network, enhancing its capability to adapt to changing circumstances.

Additionally, strategy diversity can prevent the network from being trapped in local optima, where all individuals adopt the same strategy without considering alternative approaches. Such homogeneity can hinder the network's ability to respond to new challenges and opportunities.

Real-World Applications

Network equilibrium and strategy evolution have practical implications across various domains, including social media, marketing, and public health. Let's explore some real-world examples where these concepts come into play.

1. **Social Media Influence** In social media platforms like Twitter or Instagram, network equilibrium and strategy evolution determine the popularity and adoption of certain trends or hashtags. Users observe

the strategies employed by influential individuals and imitate them to gain social capital and visibility within the network.

Understanding the dynamics of network equilibrium can help marketers and social media influencers strategically position themselves to maximize their impact. By analyzing the structure of the network and the strategies adopted by influential users, they can adapt their own strategies to reach a larger audience.

2. **Viral Marketing Campaigns** In the realm of marketing, the success of viral marketing campaigns relies on the principles of network equilibrium and strategy evolution. Marketers aim to create content that resonates with individuals and motivates them to share it within their social networks.

By leveraging social learning and evolutionary dynamics, marketers can design campaigns that tap into the collective behavior of individuals within the network. They analyze the strategies that have gained traction in the past and adapt their content to align with those successful strategies, increasing the chances of virality.

3. **Disease Spread and Vaccination Strategies** Network equilibrium and strategy evolution also play a crucial role in understanding the spread of infectious diseases and designing effective vaccination strategies. In a network of individuals, the extent of connections and interactions shapes the transmission pathways of diseases.

Analyzing network equilibrium helps epidemiologists identify central individuals who are more likely to transmit diseases and target them for intervention strategies. Moreover, understanding the dynamics of strategy evolution guides the design of vaccination campaigns, considering factors such as social influence and patterns of behavior adoption.

By addressing the diversity of strategies within a network, public health interventions can better accommodate different beliefs and attitudes towards vaccination, increasing overall vaccine uptake.

Further Resources and Challenges

If you want to explore network equilibrium and strategy evolution in more depth, a great starting point is the work of Matthew O. Jackson, a renowned researcher in social and economic networks. His book,

"Social and Economic Networks," provides a comprehensive overview of network theory and its applications.

One of the challenges in studying network equilibrium and strategy evolution is the complexity of real-world networks. Social networks can be highly dynamic, making it difficult to capture the full range of interactions and strategies employed by individuals. However, researchers are continuously developing new models and computational techniques to tackle these challenges and gain a better understanding of network dynamics.

Chapter Summary

In this section, we explored the concept of network equilibrium and how strategies evolve within social networks. We discussed the interplay between strategy and centrality, the role of social learning and evolutionary dynamics, and the importance of strategy diversity within networks.

Furthermore, we examined real-world applications of network equilibrium and strategy evolution in domains such as social media, marketing, and public health. Understanding these dynamics can offer valuable insights into human behavior, decision-making processes, and the diffusion of information and innovation within networks.

In the next chapter, we will dive into the fascinating realm of evolutionary biology and explore how game theory helps us understand the evolution of strategies in biological systems.

Diffusion and Adoption in Social Networks

In the interconnected world of social networks, information and ideas spread like wildfire. Whether it's a viral meme, a popular trend, or a new technological innovation, the process of diffusion and adoption shapes the dynamics of social networks. In this section, we will explore how game theory can help us understand and analyze the mechanisms behind diffusion and adoption in social networks.

Understanding Diffusion

Diffusion refers to the spread of an idea, behavior, or innovation through a social network. It is a process that involves the adoption and transmission of information from one individual to another.

Understanding how diffusion occurs in social networks is crucial for marketers, policymakers, and researchers alike.

One way to study diffusion is through the lens of game theory. Game theory provides a framework to analyze how individuals make decisions in a strategic setting. In the context of diffusion, we can view the process as a game where individuals are the players and their decisions to adopt or reject the idea are their strategies.

The Diffusion Game

To model the diffusion game, let's consider a simple scenario. Suppose there are two individuals, Alice and Bob, in a social network. Alice has already adopted a new idea or product, while Bob has not. Each individual faces a decision: to adopt or not to adopt.

We can represent this game using a payoff matrix:

	Bob's Strategy
Alice's Strategy	(A, B)
	(A', B')

Here, (A, B) represents the payoff to Alice and Bob if both adopt, (A', B') represents the payoff if neither adopt, and the payoffs are defined such that $A > A'$ and $B > B'$.

Strategies in Diffusion

In the diffusion game, players have two main strategies: adopt or reject. The outcome of the game depends on the individual's decision as well as the decision of others in the network.

When faced with the decision to adopt or reject, individuals consider several factors: perceived benefits, costs, social influence, and network structure. They weigh the potential rewards of adoption against the risks and costs involved. Decisions may also be influenced by individuals' social ties and their position within the network.

Diffusion Models

Game theory helps us understand diffusion in social networks by providing various models and frameworks. One commonly used model is the Bass diffusion model, which combines elements of game theory and epidemiology. The Bass model assumes that adoption occurs in two stages: innovation and imitation. Individuals are divided into two categories: innovators and imitators. Innovators adopt the new idea first, and their adoption influences imitators to follow suit.

Another popular model is the threshold model, which considers the influence of social ties within the network. In this model, individuals have a certain threshold level of adoption required to be influenced. If enough of their immediate connections adopt, they too are likely to adopt.

Facilitating Diffusion

Marketers and policymakers often aim to facilitate diffusion to promote the adoption of ideas, products, or behaviors. Understanding the dynamics of diffusion in social networks can help them design strategies to optimize the process.

Here are a few strategies commonly used to facilitate diffusion:

- **Seeding:** Identifying influential individuals or "seeds" within the network who are likely to adopt and spread the idea. By targeting these individuals, marketers can kickstart the diffusion process.

- **Social Proof:** Leveraging social influence and testimonials to encourage adoption. When individuals see others in their social network adopting the idea, they are more likely to follow suit.

- **Incentives:** Providing incentives or rewards for adoption. This can be in the form of discounts, exclusive access, or other benefits that motivate individuals to adopt.

- **Network Analysis:** Analyzing the structure of the social network to identify key connectors and communities. By understanding the network's topology, marketers can target their efforts more effectively.

Real-World Applications

Diffusion and adoption in social networks have extensive real-world applications. For example:

- **Marketing:** Understanding how ideas and trends spread in social networks helps marketers optimize their advertising and promotion strategies.

- **Public Health:** Studying diffusion is crucial in promoting public health behaviors such as vaccination and healthy habits. By identifying influential individuals within communities, public health interventions can be targeted effectively.

- **Technology Adoption:** The adoption of new technologies, such as social media platforms or mobile apps, often follows patterns of diffusion in social networks. Understanding these patterns helps technology companies design strategies for user acquisition and retention.

Conclusion

Diffusion and adoption in social networks are complex processes influenced by individual decisions, social influence, and network structure. Game theory provides us with valuable tools to analyze and understand these processes. By applying game theory to the study of diffusion, we can gain insights into how ideas spread, and use those insights to design effective strategies in various domains, from marketing to public health. Whether it's the adoption of a new product or the spread of a social movement, understanding the game of diffusion helps us navigate the dynamics of social networks in the digital age.

Exercise: Think of a recent trend or idea that has gained significant popularity on social media. Using game theory concepts, analyze the factors that led to its diffusion in the network. What strategies were employed to facilitate its adoption? How did social influence and network structure play a role in its spread?

Cooperation and Trust in Social Networks

In this section, we will explore the fascinating relationship between cooperation and trust within social networks. Social networks have become an integral part of our daily lives, connecting people from all walks of life and providing platforms for interaction and collaboration. Understanding how cooperation and trust emerge and evolve within these networks is essential for grasping the dynamics of social behavior and decision-making.

The Importance of Cooperation

Cooperation lies at the heart of social networks. It refers to individuals working together towards a common goal, even if their individual interests may conflict. Cooperation can take various forms, ranging from simple acts of kindness to complex collaborations.

In social networks, cooperation is crucial for the functioning of various systems and processes. For instance, in online communities, cooperation enables the sharing of knowledge, the development of new ideas, and the creation of valuable resources. In economic networks, cooperation drives trade, investment, and the establishment of trust among market participants. Cooperation also plays a significant role in decision-making processes within social networks, impacting how individuals interact and influence each other.

The Challenge of Cooperation

Cooperation, however, is not always guaranteed within social networks. Individuals may face dilemmas where the temptation to act in their own self-interest can undermine collective goals.

One classic example is the Prisoner's Dilemma. In this scenario, two individuals are arrested for a crime but have no evidence against each other. The prosecutor offers each prisoner a deal: if one remains silent while the other confesses, the confessor will receive a reduced sentence, while the silent one faces severe penalties. If both remain silent, both receive a moderate sentence. The dilemma arises when each prisoner must decide whether to trust the other to remain silent or confess to minimize their own sentence.

The Prisoner's Dilemma captures the tension between self-interest and cooperation, illustrating the challenges individuals face when trying to achieve collective goals within social networks.

Social Networks as Cooperation Platforms

Social networks provide a fertile ground for the emergence of cooperation. The structure of social connections can influence individuals' decisions to cooperate or defect. Through social interactions, individuals may observe and learn from others' behaviors, adjust their strategies, and establish trust.

One influential concept in social networks is social capital. Social capital refers to the resources embedded in social networks, including trust, reciprocity, and cooperation. People with high social capital tend to have more connections, stronger relationships, and greater access to resources. This builds a foundation for cooperation and collaboration, as individuals can rely on trusted peers within their network.

Trust in Social Networks

Trust is a fundamental element in promoting cooperation within social networks. Trust refers to the belief that others will act in a reliable and cooperative manner, even in uncertain situations. Trust allows individuals to engage in cooperative actions without fear of being exploited or betrayed.

Building trust within a social network is a complex process. It requires individuals to assess the trustworthiness of others based on their reputation, past behaviors, and social norms. A person with a good reputation for honesty and fairness is more likely to be trusted than someone with a history of deception.

Trust can also be influenced by network characteristics. Strong ties between individuals, such as close friendships or family relationships, tend to foster higher levels of trust. This is because close ties often involve frequent interactions, shared experiences, and mutual obligations, which contribute to a sense of trust and cooperation.

Maintaining Trust in Social Networks

While trust can be formed, it can also be easily broken within social networks. The maintenance of trust is critical for sustaining cooperation over time. Various factors can impact the fragility or resilience of trust within a network.

One factor is the existence of social norms and institutions that encourage and enforce cooperative behavior. These norms provide a shared understanding of expected behaviors and discourage opportunistic actions. In economic networks, for example, reputable online marketplaces often employ rating systems and dispute resolution mechanisms to promote trust among buyers and sellers.

Transparency and information availability within a network also play a role in maintaining trust. When individuals have access to reliable information about others' past behavior, they can make informed decisions about whom to trust and cooperate with. In online communities, platforms that provide user ratings and reviews empower individuals to make better decisions about whom to engage with.

The Dark Side: Betrayal and Exploitation

While social networks have the potential to foster cooperation and trust, they are not immune to betrayal and exploitation. In some cases, individuals may exploit the trust of others for personal gain, causing harm and disrupting cooperative dynamics.

Understanding the dynamics of betrayal and exploitation within social networks is essential for managing the risks associated with cooperation. This involves identifying vulnerabilities, designing robust systems, and developing strategies to mitigate the impact of betrayal.

Real-World Applications

The study of cooperation and trust in social networks has significant real-world applications. It provides insights into the functioning of online communities, the dynamics of economic systems, and the formation of social norms. Understanding how cooperation and trust emerge and evolve within social networks can help us design better systems, establish fairer economic structures, and promote positive social interactions.

For example, online platforms can employ game-theoretic models and algorithms to incentivize cooperation and discourage defection. By rewarding users for trustworthy behavior and penalizing those who exploit others' trust, platforms can create an environment where cooperation thrives.

In the realm of international relations, understanding trust-building mechanisms within social networks can facilitate conflict resolution and negotiation processes. By identifying key actors and developing strategies to establish trust, diplomats and negotiators can work towards sustainable solutions.

Summary

Cooperation and trust play critical roles in shaping social networks. Despite the challenges posed by self-interest and the potential for betrayal, social networks have proven to be fertile grounds for the emergence of cooperation. Understanding the dynamics of cooperation and trust within social networks has broad implications for fields ranging from economics and politics to psychology and sociology.

By studying the factors that influence cooperation and trust, we can develop strategies to foster positive social interactions, build stronger communities, and create more equitable economic systems. The study of cooperation and trust in social networks is a vital area of research that offers valuable insights into human behavior and decision-making.

Game Theoretic Models of Opinion Formation

In this section, we will explore how game theory can be used to model opinion formation in social networks. Human beings are highly influenced by the opinions and behaviors of those around them, and understanding how opinions spread and evolve is crucial in various fields, including sociology, political science, marketing, and public opinion research. Game theory provides a powerful framework to analyze the dynamics of opinion formation and the strategies individuals adopt in response to others' opinions.

Social Influence and Opinion Dynamics

Opinions are not formed in isolation; individuals are influenced by the opinions and behaviors of others. Social influence plays a vital role in shaping our attitudes and beliefs, and understanding how this influence occurs is a fundamental aspect of opinion formation.

One widely used model for opinion dynamics is the DeGroot model, which assumes that individuals update their opinions by taking a weighted average of the opinions of others in their social network. The weights reflect the influence that individuals attribute to their neighbors. This model captures the idea that individuals trust and value the opinions of others to varying degrees.

Formally, let $G = (N, E)$ be a social network, where N is the set of individuals and E is the set of edges representing social connections. Each individual i has an opinion denoted by x_i, and the weight attributed to the opinion of individual j by individual i is denoted by w_{ij}. The DeGroot model can then be described by the following equation:

$$x_i(t+1) = \sum_{j \in N} w_{ij} x_j(t)$$

This equation states that the updated opinion of individual i at time $t+1$ is the weighted sum of the opinions of their neighbors at time t. The process iterates until convergence, where opinions stabilize.

Opinion Cascades and the Voter Model

Opinion cascades occur when individuals adopt the opinions of others in a sequential manner. These cascades can result in the rapid spread of opinions within a social network. The voter model is a simple yet powerful model to study opinion cascades.

In the voter model, each individual has one of two opinions, denoted as 0 or 1. At each time step, an individual is selected at random, and they update their opinion to match one of their neighbors' opinions with a certain probability. This updating process continues until a consensus is reached, where all individuals hold the same opinion.

The voter model can provide insights into the dynamics of opinion cascades and the conditions under which consensus is reached. It demonstrates that the density of connections within a network and the initial distribution of opinions play a crucial role in determining the final outcome.

Game-Theoretic Approaches

Game-theoretic approaches to opinion formation consider individuals as strategic actors who aim to maximize their own utility. These approaches introduce the concept of strategic interactions, where individuals take into account the opinions of others and the potential benefits or costs associated with a particular opinion.

One widely studied game-theoretic model is the binary voter model with persuasion. In this model, individuals not only update their opinions based on their neighbors' opinions but also consider the persuasiveness of the arguments supporting different opinions. Each individual assigns a level of persuasiveness to each opinion, and based on these evaluations, they update their opinions.

Formally, let N be the set of individuals, and x_i be the opinion of individual i, where $x_i \in \{0, 1\}$. Let A be the set of possible opinions, and s_{ij} be the persuasiveness of opinion j for individual i, where $s_{ij} \geq 0$. The individual i's updated opinion, denoted as x'_i, is given by:

$$x'_i = \arg\max_{a \in A} \sum_{j \in N} s_{ij} \mathbb{1}_{x_j = a}$$

This equation states that individual i selects the opinion that maximizes the sum of the persuasiveness weights of the arguments supporting that opinion.

The binary voter model with persuasion captures the strategic aspect of opinion formation, where individuals not only react to others' opinions but also consider the strength of the arguments supporting those opinions. This model highlights the role of persuasion and the importance of well-constructed arguments in influencing others' opinions.

Applications and Real-World Examples

Game theoretic models of opinion formation find numerous applications in various domains. In marketing, these models can be used to understand how opinion leaders and influencers shape consumers' attitudes towards products and brands. By identifying key individuals with high persuasiveness scores, marketers can strategically target these individuals to amplify positive opinions and minimize negative ones.

In political science, game-theoretic models help us understand the dynamics of elections and coalition formation. By modeling opinion formation as a strategic process, these models shed light on the strategies political parties adopt to attract voters and form alliances. Such insights can inform campaign strategies and policy-making processes.

Game-theoretic models also have implications in social media analysis, where understanding opinion dynamics is crucial. By studying how opinions spread on platforms like Twitter or Facebook, we can better understand the mechanisms behind the formation of echo chambers and filter bubbles. This knowledge can inform technology design and policy interventions to mitigate the negative effects of polarization.

Conclusion

Game-theoretic models provide a powerful framework for analyzing and understanding opinion formation in social networks. These models capture the dynamics of social influence, opinion cascades, and strategic interactions. By studying how opinions evolve and spread, we can gain insights into collective decision-making processes, marketing strategies, and political dynamics. Through a combination of mathematical modeling and empirical analysis, game theory offers valuable tools for understanding and predicting how opinions form and change in our interconnected world.

Real-World Applications of Game Theory in Social Networks

Social networks have become an integral part of our lives, shaping the way we connect, interact, and share information. From Facebook to Twitter to Instagram, these platforms have revolutionized the way we communicate and form relationships. But what if I told you that game theory can help us understand and analyze these social networks? Yeah, you heard me right. Game theory isn't just for nerds sitting in a classroom; it can be applied to real-world situations, like social networks. So buckle up, because we're about to dive into some fascinating applications of game theory in social networks.

Network Structures and Behavior

Let's start by understanding the structure of social networks. They can be represented as graphs, where individuals are nodes, and their connections are edges. The behavior of individuals in a social network is influenced by their connections, and game theory provides tools to study this behavior. One concept that comes in handy when analyzing social networks is centrality.

Centrality and Influence in Social Networks

Centrality refers to the importance or influence of a particular individual within a social network. It measures how well-connected and influential someone is in the network. There are different measures of centrality, such as degree centrality, closeness centrality, and betweenness centrality.

Degree centrality measures the number of connections an individual has. In a social network, a person with high degree centrality would be someone with a large number of friends or followers. They tend to have more influence, as their actions or opinions can reach a wider audience.

Closeness centrality measures how quickly an individual can reach all other individuals in the network. Someone with high closeness centrality can access information or influence others more efficiently. For example, if you have a friend who always seems to be the first to know about the latest gossip, they likely have high closeness centrality.

Betweenness centrality measures how often an individual lies on the shortest path connecting other individuals in the network. Individuals with high betweenness centrality act as bridges or intermediaries between different groups or clusters within the network. They have the potential to control information flow or broker connections.

Understanding centrality is crucial because it helps us identify key players and influencers in social networks. By applying game theory, we can analyze how these influential individuals strategically use their power and connections to achieve their goals.

Small World Phenomenon and Social Connectivity

Ever heard of the "six degrees of separation" concept? It suggests that any two people in the world are connected by, at most, six degrees of separation. In other words, you can reach anyone on the planet by a chain of personal connections consisting of no more than six individuals. This phenomenon is known as the small world phenomenon.

Game theory allows us to study how information spreads and travels through social networks, contributing to the small world phenomenon. Researchers use game theory models to understand how individuals make decisions about who to connect with and share information with. For example, the famous "prisoner's dilemma" can be applied to social networks to analyze how trust and cooperation influence the spread of information.

Homophily and Social Influence

Have you ever noticed how your friends tend to have similar interests, tastes, or behaviors? This phenomenon is known as homophily. Game theory provides insight into how homophily affects social influence and decision-making within networks.

Homophily can create echo chambers, where individuals only interact with like-minded people and reinforce their existing beliefs or opinions. Game theory models help us understand how these echo chambers form and what impact they have on the spread of information.

Game theorists also study how individuals can leverage social influence within their network. We often rely on recommendations or opinions from our friends when making decisions. By understanding the dynamics of social influence, game theory can guide us in designing strategies to maximize the impact of our messages or actions within a social network.

Network Dynamics and Game Theory

Social networks are not static entities; they evolve and change over time. Game theory plays a crucial role in studying the dynamics of social networks by analyzing how individuals adapt their strategies based on the actions and behaviors of others.

Models such as evolutionary game theory help us understand how certain behaviors or strategies become dominant within a social network. We can study how cooperation, trust, and conflict emerge and evolve within these networks. For example, by using game theory, researchers have explored how the emergence of cooperation can be influenced by factors like reputation and punishment.

Network Equilibrium and Strategy Evolution

In social networks, individuals often face the challenge of selecting strategies that maximize their own benefits while considering the actions of others. Game theory provides insights into the equilibrium points that emerge in social networks, where no individual can unilaterally deviate from their chosen strategy to improve their outcomes.

Understanding network equilibrium is crucial for predicting the behavior of individuals in a social network. Game theory allows us to analyze how strategies evolve over time as individuals imitate or learn from each other. By studying the dynamics of strategy evolution, game theorists can make predictions about the future behavior of individuals within a network.

Diffusion and Adoption in Social Networks

One of the key applications of game theory in social networks is the study of diffusion and adoption processes. Game theory models help us understand how information, innovations, or behaviors spread within a social network and how they are adopted by individuals.

For example, the "viral marketing" phenomenon can be studied using game theory. By analyzing the incentives and strategies of individuals within a network, we can determine how to maximize the spread and adoption of a product or idea.

Understanding the diffusion and adoption processes in social networks can have practical applications in fields like marketing, public health, and policy-making. By identifying key influencers and understanding the factors that drive adoption, we can design effective strategies to promote positive behavior change or increase the uptake of innovations.

Cooperation and Trust in Social Networks

Cooperation and trust are essential for the smooth functioning of any social network. Game theory provides valuable insights into the mechanisms and conditions that foster cooperation and trust within social networks.

One widely studied game in the context of cooperation is the "prisoner's dilemma." In this game, two individuals must decide whether to cooperate or defect, without knowing the other's decision. By studying the strategies and outcomes of repeated prisoner's dilemma games within a network, game theorists can uncover the conditions that lead to cooperation or defection.

Game theory also helps us understand how trust is built and maintained in social networks. Models like the "trust game" capture the dynamics of trust by examining how individuals make decisions when faced with uncertainty about others' intentions.

Understanding cooperation and trust within social networks has broad implications, from fostering collaboration in online communities to building trust in online transactions or peer-to-peer platforms.

Game Theoretic Models of Opinion Formation

Opinions and beliefs play a central role in social networks, shaping interactions and influencing decision-making. Game theory provides models that explain how opinions form and spread within a network.

Opinion dynamics models in game theory capture the interplay between individuals' beliefs, their interactions, and the influence exerted by others. By studying these models, researchers gain insights into phenomena like opinion polarization, the emergence of consensus, and the role of influencers in opinion formation.

Understanding how opinions form and spread within social networks has applications in various areas, including political science, marketing, and public opinion research.

Real-World Applications of Game Theory in Social Networks

The applications of game theory in social networks are vast and continue to grow as our understanding of these networks deepens. Some real-world applications include:

- Social media strategies: Companies can leverage game theory to design effective social media marketing campaigns, targeting influential individuals and optimizing message diffusion.

- Online reputation management: Game theory models can help individuals and organizations manage their online reputation by understanding the dynamics of influence and trust within social networks.

- Disease spreading: Game theory is used to study the spread of diseases within social networks, guiding interventions and preventive strategies.

- Online communities and collaboration: Game theory can help design systems that encourage cooperation and collaboration within online communities, such as open-source software development or crowd-based problem-solving platforms.

- Online advertising: By understanding the dynamics of attention and influence within social networks, game theory can inform the design of targeted online advertising strategies.

- Social network analysis: Game theory is a valuable tool for analyzing social networks, uncovering patterns of behavior, identifying key actors, and predicting their actions.

As you can see, game theory has a lot to offer when it comes to understanding and analyzing social networks. By applying game theory principles to real-world social network scenarios, we gain insights into the complexities of human behavior, cooperation, influence, and decision-making. So, the next time you check your social media feed, remember that there's more to it than meets the eye, and game theory is there to help us unravel its secrets.

Now that we've explored the applications of game theory in social networks, we'll turn our attention to another fascinating intersection:

game theory and evolutionary biology. Get ready to explore how game theory sheds light on the complexities of natural selection, evolution, and animal behavior.

Game Theory and Evolutionary Biology

Evolutionary Game Theory

Darwinian Evolution and Game Theory

Darwinian evolution, proposed by the legendary naturalist Charles Darwin, revolutionized our understanding of how life on Earth has evolved over time. It introduced the concept of natural selection as the driving force behind the adaption and diversification of species. In the context of game theory, Darwinian evolution plays a crucial role in understanding the dynamics of strategic interactions and the emergence of cooperation and competition among organisms.

Principles of Darwinian Evolution

At the heart of Darwinian evolution lies the principle of natural selection. According to this principle, individuals within a population exhibit variation in traits, and these variations impact their chances of survival and reproduction. Traits that enhance an organism's fitness and improve their ability to adapt to their environment are more likely to be passed on to future generations. Over time, this process leads to the accumulation of advantageous traits and the divergence of species.

The concept of fitness, central to Darwinian evolution, is a measure of an organism's reproductive success. Fitness is determined by the ability of an organism to survive and reproduce, passing on its genetic material to subsequent generations. The fittest individuals are those with traits that increase their likelihood of survival and successful reproduction.

Evolutionary Game Theory

Evolutionary game theory is a powerful tool used to analyze and understand how strategic interactions among individuals in a population influence the evolution of traits and behaviors. It combines the principles of game theory with the principles of Darwinian evolution to model the dynamics of evolutionary processes.

In game theory, a game consists of players, strategies, and payoffs. Similarly, in evolutionary game theory, a population consists of individuals, strategies, and fitness payoffs. Individuals with different strategies compete for resources, mates, or other benefits, and their relative fitness determines the proportion of strategies in the next generation.

To model the interactions between individuals in a population, evolutionary game theory employs the concept of an evolutionary stable strategy (ESS). An ESS is a strategy that, once it becomes prevalent in a population, cannot be invaded by any other mutant strategy. In other words, an ESS is a strategy that, if adopted by the majority, ensures stability and prevents the population from being invaded by alternative strategies.

The Prisoner's Dilemma in Evolutionary Biology

One of the most famous games in game theory, the Prisoner's Dilemma, also has significant implications for evolutionary biology. The Prisoner's Dilemma captures a fundamental tension between cooperation and defection. In this game, two individuals have the choice to either cooperate with each other for mutual benefit or defect and maximize their own benefits at the expense of the other.

Translating the Prisoner's Dilemma into the context of evolutionary biology, cooperation may involve behaviors such as sharing resources, defending against predators, or cooperating in raising offspring. Defection, on the other hand, may involve exploiting others' cooperative behaviors without reciprocating.

Cooperation can bring about mutual benefits, but it is vulnerable to exploitation by defectors. If cooperation becomes prevalent in a population, defectors can exploit the cooperators, reaping the benefits without incurring the costs of cooperation. This leads to a selection pressure favoring defectors, potentially driving cooperators to extinction.

Evolutionary Dynamics of Cooperation

The study of cooperation in evolutionary game theory aims to understand the conditions under which cooperation can emerge and thrive in the face of defectors. One key mechanism that promotes cooperation is spatial structure, where individuals interact primarily with their neighbors in a structured population.

In a spatially structured population, cooperators can form clusters, promoting the evolution and maintenance of cooperation. Defectors, however, face the risk of being surrounded by cooperators, reducing their fitness compared to cooperators within clusters. This creates an evolutionary advantage for cooperation in spatially structured populations.

Furthermore, different strategies can coexist in a population through the mechanism of frequency-dependent selection. In frequency-dependent selection, the relative fitness of a strategy depends on its frequency in the population. If a strategy becomes common, its fitness advantage diminishes, and alternative strategies can exploit the common strategy's weaknesses, creating an evolutionary equilibrium with multiple strategies coexisting.

Applications of Evolutionary Game Theory

Evolutionary game theory has found applications in various fields, including ecology, sociology, and economics. It can shed light on the evolution of cooperation among animals, the spread of cooperative behaviors in human societies, and the dynamics of economic interactions.

For example, studying the evolution of cooperation in animal societies can provide insights into the development of complex social structures and the management of shared resources. Understanding the emergence and stability of cooperation in human societies can help address challenges such as climate change, resource allocation, and societal conflicts.

In economics, evolutionary game theory can inform our understanding of the dynamics of market competition, the emergence of cooperation or collusion among firms, and the evolution of consumer behavior. It offers a powerful framework to analyze strategic interactions and predict the outcomes of economic decision-making.

The Evolution of Cooperation: A Real-World Example

A well-known real-world example of cooperation is seen in the behavior of cleaner fish and their hosts in coral reef ecosystems. Cleaner fish, such as cleaner wrasses, remove parasites from the bodies of larger fish, known as the hosts. This behavior benefits both the cleaner fish, who obtain nourishment, and the hosts, who get rid of parasites.

However, the cleaner fish could choose to cheat by taking bites of healthy tissue from the hosts instead of removing parasites. This would provide immediate benefits to the cleaner fish without providing any cleaning services. If such cheating behavior became prevalent, hosts would be less likely to approach cleaner fish, leading to a breakdown in cooperation.

The evolutionary dynamics of this system have been studied using game theory. It has been found that when cleaner fish and hosts interact repeatedly, a tit-for-tat strategy can evolve, where the cleaner fish only cheats if the host previously cheated. This reciprocal behavior helps maintain cooperation between the cleaner fish and the hosts, leading to mutual benefits.

Conclusion

Darwinian evolution and game theory are intimately linked, offering a powerful framework to understand the dynamics of strategic interactions and the emergence of cooperation and competition in biological systems. By combining the principles of natural selection and game theory, evolutionary game theory provides valuable insights into the evolution of traits and behaviors in a wide range of contexts, from the behavior of animals to human societies and economic systems.

Fitness and Replicator Dynamics

In the study of evolutionary game theory, understanding how fitness and replicator dynamics interact is crucial. Fitness refers to the ability of an individual or a strategy to survive and reproduce in a given environment. Replicator dynamics, on the other hand, focuses on how the proportion of different strategies changes over time. Together, these concepts shed light on the dynamics of evolutionary games and the long-term survival of strategies within a population.

Introduction to Fitness

Fitness is a measure of the success of a strategy or an individual in a given environment. In the context of evolutionary game theory, fitness is often quantified in terms of reproductive success. Individuals or strategies with higher fitness are more likely to produce offspring, pass on their genetic material, or achieve their objectives.

The concept of fitness can be influenced by various factors, including environmental conditions, interactions with other individuals, and the specific goals of the game. For example, in a predator-prey scenario, the fitness of a predator strategy would be determined by its ability to catch prey and survive, while the fitness of a prey strategy would be determined by its ability to avoid being caught.

In mathematical terms, fitness can be represented by a fitness function, which assigns a numerical value to each strategy based on its success in achieving specific objectives. The fitness function is usually defined in such a way that strategies with higher fitness values have a higher probability of survival and reproduction.

Replicator Dynamics

Replicator dynamics is a mathematical framework used to model the evolution of strategies in a population. It provides a way to analyze how the frequencies of different strategies change over time and how the population as a whole evolves.

The key idea behind replicator dynamics is that the fitness of a strategy determines its reproductive success, which in turn affects the proportion of individuals using that strategy in the next generation. Strategies with higher fitness values will have a higher reproduction rate, leading to an increase in their frequency in the population. Conversely, strategies with lower fitness values will have a lower reproduction rate, resulting in a decrease in their frequency.

Mathematically, replicator dynamics can be described using differential equations or difference equations. The exact form of the equations depends on the specific game being studied. In general, the equations capture the change in the proportion of each strategy over time, taking into account the fitness values of the strategies and their current frequencies.

Solving the replicator dynamics equations allows us to understand the long-term behavior of strategies in an evolutionary game. This analysis can reveal stable equilibria, where the frequencies of strategies remain constant, as well as unstable equilibria, where the frequencies fluctuate or change over time.

Applying Fitness and Replicator Dynamics

The study of fitness and replicator dynamics has practical applications in various fields, including biology, economics, and sociology. Understanding how strategies evolve and compete in different environments can provide insights into the dynamics of biological systems, economic markets, and social interactions.

For example, in biology, fitness and replicator dynamics help explain the evolution of certain traits or behaviors in a population. By studying the fitness of different strategies, such as cooperative or competitive behaviors, researchers can gain a better understanding of how these traits evolve over time and why certain strategies persist or decline.

In economics, fitness and replicator dynamics are used to analyze the behavior of firms, consumers, and markets. By modeling the evolution of strategies in competitive settings, economists can study the dynamics of price competition, market entry and exit, and the emergence of new technologies or products.

In sociology, fitness and replicator dynamics provide a framework for studying the spread of ideas, opinions, and social norms within a population. By examining the fitness of different strategies, such as adopting or rejecting a new behavior, sociologists can explore the factors that influence the diffusion of innovations and the formation of social networks.

Example: Cooperation and Defection

To illustrate the interplay between fitness and replicator dynamics, let's consider a classic example of the Prisoner's Dilemma game. In this game, two players can choose to cooperate or defect, with the following payoff matrix:

	Cooperate	Defect
Cooperate	R,R	S,T
Defect	T,S	P,P

In this matrix, R represents the reward for mutual cooperation, S represents the temptation to defect, T represents the sucker's payoff, and P represents the punishment for mutual defection.

Assuming a large population of players interacting repeatedly, we can analyze the long-term dynamics of cooperation and defection using fitness and replicator dynamics. The fitness of a strategy can be quantified based on the accumulated payoff over multiple interactions.

If initially, there is a mix of cooperators and defectors in the population, the replicator dynamics equations will determine the change in the frequencies of these strategies over time. Depending on the payoff values, different equilibria may emerge.

For example, if the temptation to defect (S) is high relative to the reward for mutual cooperation (R), the population may converge to an equilibrium where defection dominates. This is because defectors have a higher fitness and reproductive success in this scenario.

Alternatively, if the temptation to defect is low and the reward for mutual cooperation is high, cooperation may prevail in the long run. In this case, cooperators have a higher fitness and are more likely to reproduce, resulting in a higher frequency of cooperation in the population.

The interplay between fitness and replicator dynamics in this example illustrates how the success or failure of strategies can shape the evolution of cooperation and defection. Understanding these dynamics can provide insights into real-world scenarios where cooperation is essential, such as collective action problems, social dilemmas, and the evolution of altruistic behavior.

Conclusion

Fitness and replicator dynamics play a crucial role in understanding the dynamics of evolutionary games. By quantifying the success of strategies and modeling their evolution over time, these concepts shed light on the survival and prevalence of different strategies in a population.

The application of fitness and replicator dynamics extends beyond the realm of biology; it has practical implications in economics, sociology, and other social sciences. By studying how strategies evolve and compete in various environments, researchers can gain valuable insights into the dynamics of complex systems, decision-making processes, and the emergence of cooperative or competitive behavior.

As we delve deeper into the fascinating world of game theory, we will continue to explore the intricate interplay between fitness, replicator dynamics, and other concepts. By doing so, we can gain a deeper understanding of strategic interactions, decision making, and the dynamics of complex systems.

Recommended Resources

- Nowak, M. A. (2006). *Evolutionary Dynamics: Exploring the Equations of Life*. Harvard University Press.

- Hofbauer, J., & Sigmund, K. (1998). *Evolutionary Games and Population Dynamics*. Cambridge University Press.

- Smith, J. M. (1982). *Evolution and the Theory of Games*. Cambridge University Press.

- Gintis, H. (2009). *Game Theory Evolving: A Problem-Centered Introduction to Modeling Strategic Interaction*. Princeton University Press.

Evolutionary Stable Strategies: Surviving in a Changing World

Evolutionary Stable Strategies (ESS) is a key concept in evolutionary game theory that helps organisms survive in a changing environment. In this section, we will explore the principles of ESS, its applications in biology, and how it relates to the overall theme of Game Theory.

Background

Evolutionary game theory combines concepts from biology and game theory to study the evolution of strategies in populations of organisms. In this context, a *strategy* is a set of rules or behaviors that an organism employs in order to maximize its fitness (or reproductive success). By adapting and evolving strategies over time, organisms can better survive and reproduce in their environments.

EVOLUTIONARY GAME THEORY

In the context of ESS, a strategy is considered evolutionarily stable if it resists invasion by other strategies. In other words, once a strategy becomes established in a population, it is advantageous for other individuals to adopt the same strategy rather than trying to introduce a new one. ESS provides a stable equilibrium, ensuring the survival of the strategy in the long run.

Principles of ESS

To understand ESS, we must first examine the concept of fitness in evolutionary biology. Fitness represents an organism's ability to survive and reproduce, thereby passing on its genes to subsequent generations. Fitness depends on the interaction between strategies and the environment in which the organisms live.

In an evolutionary game, strategies compete with each other for resources or reproductive opportunities. The fitness of a strategy depends on how well it performs in this competition. The principle of ESS can be summarized as follows:

An ESS strategy is one that, when adopted by a population, cannot be invaded by an alternative strategy, given that the population is at equilibrium.

To understand why an ESS strategy is resistant to invasion, we can explore two key components: payoff and stability.

Payoff: The payoff of a strategy is the fitness or reproductive success it yields in interaction with other strategies. An ESS strategy must have a higher payoff than any other strategy competing against it. In other words, it should be capable of outperforming alternative strategies in the long run.

Stability: An ESS strategy is stable in the sense that if a small number of individuals deviate from it and adopt a different strategy, they will experience lower fitness compared to those adhering to the ESS strategy. This discourages further adoption of the deviating strategy, reinforcing the stability of the ESS.

The combination of higher payoff and stability ensures that an ESS strategy will remain dominant in a population, even in the face of potential new strategies trying to invade.

Examples of ESS

One classic example of ESS is the hawk-dove game, which illustrates the evolution of aggressive and non-aggressive strategies in animal conflicts. In this game, two individuals engage in a contest over a resource, such as mates or food. They can choose to be aggressive (hawk) or non-aggressive (dove).

If two hawks meet, they engage in a fierce battle, resulting in a lower payoff for both individuals due to injuries and expended energy. If a dove encounters a hawk, the dove retreats, yielding the resource to the hawk without a fight. When two doves interact, they peacefully share the resource, resulting in a moderate payoff for both.

The payoff matrix for the hawk-dove game can be represented as follows:

	Hawk	Dove
Hawk	a/2, a/2	0, a
Dove	a, 0	a/2, a/2

In this matrix, a represents the benefit of winning a resource, and the fractions represent the costs associated with engaging in a fight or retreating.

The ESS for this game is a mixed strategy, where individuals adopt both hawk and dove behaviors randomly. With the appropriate distribution of strategies, it becomes impossible for a new strategy, such as being always hawkish, to invade and take over the population. The ESS maintains stability by preventing any single strategy from dominating completely.

Applications of ESS

ESS has important applications in various fields, including biology, economics, and sociology. By understanding the principles of ESS, researchers can model and predict the evolution of strategies in different contexts.

In biology, ESS helps explain the evolution of behaviors such as cooperation, altruism, and territoriality. For example, the ESS concept can shed light on why animals cooperate to protect their territories and resources, even though cooperation may entail some costs.

In economics, ESS is relevant to the study of competition, pricing, and market dynamics. ESS models help analyze how firms strategize and make decisions in competitive environments. By identifying the evolutionarily stable strategies in markets, economists can better understand market dynamics and outcomes.

In sociology, ESS is useful for studying social norms and cultural evolution. It provides insights into the stability and persistence of certain norms or behaviors within societies. ESS models help analyze how individuals conform to social norms and how norms evolve over time.

Caveats and Challenges

While ESS is a powerful concept in evolutionary game theory, it has some limitations and challenges. One challenge is the assumption of perfect rationality and full knowledge of the game. In reality, individuals often have limited information and cognitive biases that can impact their decision-making.

Additionally, ESS models assume a static environment, which may not be realistic in many cases. Environments are subject to change, and strategies that were once stable may become vulnerable to invasion under new conditions. Adaptation and the ability to adjust strategies are essential for survival in a changing world.

Another caveat is that ESS does not necessarily guarantee the optimality of a strategy. It merely ensures stability in a given context. The success of a strategy may depend on the specific conditions and the interactions between strategies.

Conclusion

Evolutionary Stable Strategies provide valuable insights into the dynamics of strategy evolution in a wide range of contexts. By understanding the principles of ESS, we can better grasp the survival and reproductive success of strategies in a changing world. ESS helps us analyze and predict behaviors in biology, economics, sociology, and beyond. However, it is crucial to consider the assumptions and limitations of ESS models when applying them to real-world scenarios. In an ever-evolving landscape, flexibility and adaptability are key to thriving and surviving.

The Prisoner's Dilemma in Evolutionary Biology

In the field of evolutionary biology, game theory is a powerful tool for understanding the dynamics of interactions among individuals within a population. One of the most famous and widely studied games in evolutionary biology is the Prisoner's Dilemma.

The Prisoner's Dilemma is a two-player game that represents a situation where individuals are faced with a choice between cooperation and defection. In this game, each player can choose to cooperate with the other player or defect. The outcome of the game is determined by the combined choices of both players.

Let's consider a scenario where two individuals, Alice and Bob, are members of a population of animals. They are each given the opportunity to either cooperate by engaging in a mutually beneficial behavior, such as sharing resources, or defect by not cooperating and taking advantage of the other's cooperation.

The payoff matrix for the Prisoner's Dilemma is as follows:

	Cooperate	Defect
Cooperate	R / R	S / T
Defect	T / S	P / P

In the payoff matrix, R represents the reward for mutual cooperation, S represents the sucker's payoff (the cost of cooperating while the other player defects), T represents the temptation to defect (the benefit of defecting while the other player cooperates), and P represents the punishment for mutual defection.

In evolutionary biology, the Prisoner's Dilemma is often used to model interactions where individuals have a choice between cooperating to benefit the group or defecting to selfishly benefit themselves. The evolutionary success of different strategies can be evaluated based on the reproductive fitness of individuals.

To analyze the Prisoner's Dilemma in evolutionary biology, various strategies can be employed. One of the most well-known strategies is the Tit-for-Tat strategy. Tit-for-Tat is a cooperative strategy that starts by cooperating and then replicates the opponent's previous move at each subsequent round. This strategy promotes cooperation by reciprocating both cooperation and defection. However, it is also forgiving, as it gives the opponent a chance to revert to cooperation.

The success of different strategies in the Prisoner's Dilemma depends on the frequency of interaction, the population structure, and the presence of other strategies. Evolutionary game theorists have conducted extensive research to understand the dynamics of this game in different scenarios.

One notable result in evolutionary biology is the emergence of cooperative behavior in populations even when the dominant strategy is to defect. This is known as the evolution of cooperation. Through repeated interactions and the introduction of strategies like Tit-for-Tat, cooperation can become a stable and successful strategy, leading to higher overall fitness in the population.

The Prisoner's Dilemma has been used to study a wide range of biological phenomena, including the evolution of cooperation in social insects, the spread of selfish genes, the development of cooperative behaviors in animals, and the evolution of human cooperation and altruism.

For example, the Prisoner's Dilemma has been used to study the evolution of cooperation in bacterial populations. It has been observed that bacteria can secrete molecules that allow them to cooperate and form biofilms, which provide benefits to the entire population. However, there are always cheater bacteria that do not secrete these molecules and take advantage of the cooperation without contributing themselves. The interaction between the cooperators and cheaters mirrors the dynamics of the Prisoner's Dilemma, and studying this game helps us understand the strategies that promote cooperation in bacterial populations.

The Prisoner's Dilemma in evolutionary biology highlights the complex interplay between cooperation and selfishness in nature. It provides insights into the evolution of social behaviors and helps us understand how cooperation can emerge and persist in diverse biological systems.

Caveat: It is important to note that the Prisoner's Dilemma is a simplification of complex biological interactions. While it provides valuable insights, real-life situations often involve more intricate decision-making processes and additional factors that can influence the outcome. Evolutionary biology continues to explore and refine our understanding of the Prisoner's Dilemma and its implications for the natural world.

Further Reading:

- Nowak, M. A. (2006). *Evolutionary dynamics: exploring the equations of life*. Harvard University Press.

- Axelrod, R. (1984). *The Evolution of Cooperation*. Basic Books.

- Gardner, A., & West, S. A. (2011). The genetical theory of 'kin selection'. *Journal of Evolutionary Biology, 24*(5), 1020-1043.

Interactions between Genetic and Cultural Evolution

In this section, we will explore the fascinating interplay between genetic and cultural evolution. Both genetic and cultural factors shape our behavior, and understanding how they interact is crucial to gaining a deeper insight into human society and biology.

Genetic Evolution

Genetic evolution is the process by which favorable traits are passed down through generations. It operates on the principles of variation, inheritance, and natural selection. Genetic variation arises from mutations and recombination during reproduction. Some variations offer advantages, such as increased resistance to diseases or the ability to survive in harsh environments.

In the context of game theory, genetic evolution can influence behavior by shaping the innate predispositions and instincts of individuals. For example, in a cooperative game, individuals with a genetic predisposition for altruistic behavior may be more likely to collaborate, ultimately improving their collective fitness. On the other hand, individuals with a predisposition for selfish behavior may excel in competitive games that reward individual success.

Cultural Evolution

Cultural evolution, on the other hand, refers to the transmission and modification of knowledge, beliefs, and behaviors within a human population. Unlike genetic evolution, cultural evolution can occur rapidly, as knowledge and ideas can be shared and modified between individuals through social learning mechanisms such as imitation and teaching.

Cultural evolution allows for the accumulation of knowledge and the development of complex societies. It enables groups to adapt and learn from each other's experiences, often leading to improved strategies and outcomes. In the realm of game theory, cultural evolution shapes the strategies and norms that individuals adopt in various social interactions.

Interactions and Feedback Loops

The interactions between genetic and cultural evolution are complex and can create feedback loops that shape human societies over time. Genetic predispositions can influence the acquisition and spread of cultural traits, while cultural norms and practices can, in turn, influence the selective pressures acting on genetic traits.

For example, consider a cooperative game scenario where individuals with a genetically inherited predisposition for cooperation are more likely to succeed. This success leads to the transmission and adoption of cooperative cultural behaviors within the group, further reinforcing cooperation as a social norm. Over generations, cooperation becomes an increasingly prevalent trait within the population.

Conversely, cultural practices can also influence genetic traits. For instance, cultural norms advocating for monogamy may shape mating patterns, favoring individuals with a genetic predisposition for faithful behavior. This can influence the genetic composition of the population over time.

In some cases, the interaction between genetic and cultural evolution can lead to a co-evolutionary arms race. As individuals adapt genetically to new cultural practices, cultural practices may also evolve in response to these genetic adaptations. This dynamic interplay between the two evolutionary processes contributes to the complexity of human societies and behavior.

The Baldwin Effect: Mechanism for Genetic Assimilation

The Baldwin Effect is a concept that offers an explanation for the interaction between genetic and cultural evolution. Proposed by James Mark Baldwin in 1896, it suggests that traits initially acquired through individual learning and cultural adaptation can eventually become genetically encoded as a result of natural selection.

In the context of game theory, the Baldwin Effect can help us understand how cultural practices and strategies can eventually become innate behavioral tendencies. Initially, individuals may acquire new strategies through cultural learning and adapt their behavior accordingly. Over time, these advantageous strategies may become genetically assimilated, providing a more direct and efficient mechanism for the expression of these behaviors.

For example, consider a population that plays a competitive game where cooperative strategies yield better outcomes. Initially, individuals learn cooperative strategies through cultural transmission. However, as cooperation becomes increasingly advantageous, those individuals with a genetic predisposition for cooperative behavior may have a selective advantage. Over time, these cooperative strategies can become genetically assimilated within the population.

Case Study: The Evolution of Cooperation

One of the most prominent examples of the interplay between genetic and cultural evolution is the evolution of cooperation. Darwinian evolution, driven by genetic factors, would suggest that selfish behaviors should dominate interactions, as they provide the greatest individual fitness. However, cooperation is prevalent in human societies and can lead to mutually beneficial outcomes.

The evolution of cooperation is often explained using game theory, particularly the Prisoner's Dilemma game. This game highlights the tension between individual rationality and collective cooperation. In repeated interactions, individuals who cooperate despite short-term temptations can achieve higher long-term payoffs.

Genetic predispositions for cooperation, such as reciprocal altruism, can provide a basis for the evolution of cooperative behaviors. These predispositions can be influenced by cultural factors, such as social norms and reputational considerations. Cultural practices that promote cooperation, such as reciprocity and punishment of defectors, can further enhance the evolution of cooperation by aligning individual incentives with collective interests.

Application: Human Cooperation and Climate Change

Understanding the interplay between genetic and cultural evolution in the evolution of cooperation has significant implications for contemporary challenges, such as climate change. Overcoming the collective action problem requires individuals to cooperate and adopt sustainable behaviors.

Genetic factors, such as pro-social predispositions, may influence individuals' willingness to contribute to collective efforts to combat climate change. However, cultural factors play a crucial role in shaping individual beliefs, norms, and incentives that drive sustainable behaviors. Cultural evolution can foster the adoption of sustainable practices and incentivize cooperation on a large scale.

Efforts to mitigate climate change must consider both genetic and cultural factors. Genetic screening and understanding the genetic basis of pro-environmental behaviors can help identify individuals with a predisposition for cooperation. Additionally, cultural interventions, such as education and awareness campaigns, can promote sustainable behaviors and create social norms that align with collective goals.

Conclusion

The interplay between genetic and cultural evolution shapes our behavior, strategies, and social norms. Genetic predispositions provide a foundation for certain behaviors, while cultural evolution enables the spread and modification of behaviors within a population. Understanding the intricate relationship between these two processes is crucial for comprehending human society, biology, and the challenges we face today.

By exploring the interactions between genetic and cultural evolution, we can gain insights into how cooperation arises and how collective action problems can be addressed. This knowledge has implications for fields ranging from social sciences to biology and can inform efforts to tackle complex societal issues. The interdisciplinarity of game theory allows us to appreciate the complexity and richness of human behavior and society.

Evolutionary Game Theory in Modeling Altruism

In the study of evolutionary biology and social behavior, altruism is a fascinating and perplexing phenomenon. Altruistic behaviors involve an individual acting in a way that benefits others at a cost to itself. This seems counterintuitive from a purely selfish perspective, and yet we observe acts of kindness, cooperation, and selflessness in many species, including humans.

Evolutionary game theory provides a framework for understanding and modeling the dynamics of altruistic behaviors. It allows us to explore how natural selection can favor the emergence and maintenance of altruism, even in the face of competition.

The Evolution of Altruism

To understand the evolution of altruism, we need to examine the concepts of fitness and reproductive success. In evolutionary biology, fitness refers to an organism's ability to survive and reproduce, passing on its genes to the next generation. Natural selection acts to enhance the fitness of individuals with traits that increase their chances of reproductive success.

In the context of game theory, we can view individuals as players engaged in a game of survival. Each player's strategy represents a particular behavior, such as cooperation or defection. The payoff of a strategy is determined by its benefits and costs in terms of fitness.

Altruism, as a strategy, involves incurring a cost to provide a benefit to others. This means that altruistic individuals may experience a reduction in their own fitness due to the resources they invest in helping others. However, if the benefits of altruism outweigh the costs, altruists can still achieve a net increase in their fitness.

Kin Selection and Inclusive Fitness

One key mechanism that underlies the evolution of altruism is kin selection. Kin selection occurs when individuals help their relatives, such as siblings or cousins, because they share a portion of their genes. By helping their close relatives survive and reproduce, individuals can indirectly increase the chances of their own genes being passed on to future generations.

Kin selection relies on the concept of inclusive fitness, which combines an individual's own reproductive success with the reproductive success of its relatives. Inclusive fitness can be measured by the number of offspring an individual produces, as well as the number of offspring produced by its relatives that are directly attributable to its actions.

A well-known example of kin selection and altruism is seen in social insects like bees and ants. Workers in these colonies are predominantly female and are often sterile. They sacrifice their own reproduction to help the queen, who is their close relative, to produce more offspring. This strategy maximizes their inclusive fitness, as they contribute to the survival and reproduction of offspring that share a large portion of their genes.

Reciprocal Altruism

In addition to kin selection, reciprocal altruism is another important mechanism in the evolution of altruistic behaviors. Reciprocal altruism occurs when individuals help others with the expectation of receiving help in return at a later time. This form of cooperation is based on the premise that short-term costs can be offset by long-term benefits.

One classic example of reciprocal altruism is observed in vampire bats. These bats engage in blood-sharing behavior, where successful individuals regurgitate blood to feed less fortunate individuals that have not fed for extended periods. The reciprocal nature of this behavior ensures that bats who have been unsuccessful in finding prey on a given night are still able to survive and reproduce.

To maintain reciprocal altruism, individuals must have the ability to recognize and remember others, and to detect and punish cheaters who exploit the cooperative behavior without reciprocating. This ability to keep track of past interactions, known as a reputation system, promotes the stability of reciprocal altruism in social groups.

Game-Theoretic Models of Altruism

Game-theoretic models are used to study and analyze the dynamics of altruistic behaviors in different scenarios. In these models, individuals are represented as players who can choose between cooperative and non-cooperative strategies. The payoffs associated with these strategies depend on the interactions between individuals in a population.

One influential game-theoretic model used to study altruism is the Prisoner's Dilemma. In this two-player game, each player can choose to cooperate or defect. The payoff matrix rewards defection when the other player cooperates, but the highest payoff is achieved when both players cooperate. Despite the individual incentive to defect, cooperation can emerge and be sustained through various mechanisms, such as repeated interactions or the possibility of punishment for defection.

Another important model is the Iterated Prisoner's Dilemma, where players repeatedly face the Prisoner's Dilemma over multiple rounds. This model allows for the emergence of reciprocal altruism, as players can develop strategies that encourage cooperation while punishing defectors. The Tit-for-Tat strategy, for example, involves initially cooperating and subsequently mirroring the opponent's previous move. This strategy promotes cooperation and can deter defection.

Real-World Applications

The study of altruism and evolutionary game theory has far-reaching implications in various fields, including biology, sociology, and economics. Understanding the factors that drive altruistic behaviors can help us address pressing societal issues, such as cooperation in resource management, the emergence of social norms, and the design of efficient economic systems.

In conservation biology, for instance, game-theoretic models can shed light on the effectiveness of cooperative strategies in mitigating the depletion of shared resources. By studying the interactions between stakeholders in fisheries or deforestation, scientists and policymakers can develop strategies that promote sustainable practices and preserve ecological balance.

In the realm of economics, game theory can be used to understand phenomena like charitable giving, volunteerism, and the formation of social networks. By modeling the decision-making processes of individuals and examining the incentives for altruistic behavior, researchers can gain insights into the motivations and dynamics of social cooperation, which can inform policy interventions and the design of incentive mechanisms.

Exercise

Imagine you are a researcher studying the evolution of altruism in a particular animal population. Design a game-theoretic model that incorporates kin selection and reciprocal altruism. Consider factors such as relatedness, costs, benefits, and the potential for punishment of non-cooperative behaviors. Analyze the dynamics of cooperation in your model and discuss the conditions under which altruistic behaviors are likely to emerge and be maintained.

Game Theory and the Evolution of Cooperation

In the game of life, cooperation is a powerful force that drives social interactions and shapes the outcome of various situations. From ants working together to build colonies to humans collaborating on complex projects, cooperation has proven to be crucial for survival and success. But why do individuals choose to cooperate instead of acting solely in their own self-interest? This question lies at the heart of the study of cooperation in game theory.

Understanding Cooperation

Cooperation, in the context of game theory, refers to individuals working together to achieve a common goal, even if their immediate self-interest might suggest otherwise. It is typically explored through the lens of the Prisoner's Dilemma, a classic example in game theory.

Imagine two suspects, Alice and Bob, who are arrested for a crime. The police lack evidence for the main charge, but are confident in a lesser charge that carries a shorter sentence. The police separate Alice and Bob, and offer them both a deal: if one remains silent while the other confesses, the silent one will receive a reduced sentence, while the confessing one will go free. If both confess, they will receive a moderate sentence.

The dilemma arises from the fact that, if Alice and Bob act solely in their self-interest, both will confess and receive a moderate sentence. However, if they both choose to cooperate and remain silent, they will both receive a reduced sentence.

Evolutionary Stability and Cooperation

The evolution of cooperation has puzzled biologists and social scientists alike. How can cooperation persist when there are incentives for individuals to cheat or act selfishly?

Evolutionary game theory provides insights into this question. It suggests that cooperation can be stable and even evolve over time under certain conditions. One important concept in evolutionary game theory is the notion of an evolutionary stable strategy (ESS), which is a strategy that, when adopted by a population, cannot be invaded by alternative strategies.

In the context of cooperation, researchers have identified several mechanisms that can promote the emergence and stability of cooperative behavior. One such mechanism is direct reciprocity, where individuals interact repeatedly and have the opportunity to reward cooperators and punish defectors. By building a reputation for cooperation, individuals can forge mutually beneficial relationships and gain long-term advantages.

Tit-for-Tat and Beyond

A well-known strategy in the study of cooperation is Tit-for-Tat (TFT), which is based on the simple rule of "cooperate on the first move and then do whatever your opponent did in the previous round." TFT has been shown to be effective in promoting cooperation in repeated interactions, as it encourages cooperation, retaliates against defection, and forgives and reinitiates cooperation after a retaliatory move.

However, researchers have explored more sophisticated strategies that go beyond TFT to enhance cooperation. For example, the Generous Tit-for-Tat strategy includes occasional acts of forgiveness towards defectors, allowing them a chance to rehabilitate and reestablish cooperative interactions. Another strategy, known as Pavlov, adapts its behavior based on the outcomes of previous interactions, rewarding cooperation and punishing defection. These strategies help maintain cooperation in dynamic environments, where the payoffs and strategies of the game can change over time.

Real-World Applications

Cooperation is not only a fundamental aspect of human society, but it also plays a crucial role in various real-world applications. One such application is the study of ecological systems, where cooperation among organisms is essential for the stability and functioning of ecosystems. For example, in symbiotic relationships, such as the mutualistic interactions between flowers and pollinators, cooperation ensures the exchange of resources for survival and reproduction.

Cooperation is also relevant in economic contexts, such as in the formation of cartels or collaborations between firms. Cartels, despite being formed to manipulate prices and maximize individual profits, require cooperation among members to maintain their stability. Additionally, collaborations between firms often involve sharing resources and knowledge, which can lead to innovation and improved outcomes for all parties involved.

Understanding the dynamics of cooperation in these real-world scenarios provides practical insights that can be used to solve societal challenges. By analyzing the strategies and mechanisms that promote cooperation, we can design better policies and incentives to encourage cooperation, foster collaboration, and address collective action problems.

The Tragedy of the Commons

While cooperation has many benefits, it is not always easy to achieve. One classic example that illustrates the challenges of cooperation is the Tragedy of the Commons. This concept, originally introduced by ecologist Garrett Hardin, describes a situation where individuals, acting out of their own self-interest, deplete a shared resource, leading to its degradation or exhaustion.

Imagine a pasture that is open to all farmers for grazing their livestock. Each farmer has an incentive to graze as many animals as possible to maximize their own profit. However, if all farmers pursue this self-interest, the pasture will become overgrazed and eventually unusable. This overexploitation occurs because the cost of overgrazing is shared by all, while the benefits are captured individually.

The Tragedy of the Commons highlights the need for cooperation and collective action to manage shared resources effectively. Game theory provides tools to analyze the incentives and strategies that can counteract this tragedy and promote sustainable resource management.

In Summary

Cooperation is a fascinating aspect of human and animal behavior that can shape the outcome of various social interactions. Through the lens of game theory, we can understand the dynamics of cooperation, explore its evolutionary roots, and identify strategies that promote its emergence and stability. By applying game theory to real-world contexts, we can gain insights that help us address societal challenges and build a more cooperative and sustainable future.

So, the next time you find yourself pondering the intricacies of cooperation, remember that game theory has your back, providing a framework to unravel the mysteries behind cooperative behavior.

Now, let's dive deeper into the fascinating world of social networks and how game theory can help us navigate their complexities.

Cooperation and Cheating in Biological Systems

In the realm of biology, cooperation and cheating are fascinating phenomena that are often observed in various species. From bacteria to primates, organisms engage in cooperative behaviors to maximize their fitness and survival. However, the presence of cheaters, individuals that exploit the benefits of cooperation without contributing their fair share, can threaten the stability and sustainability of cooperative systems. In this section, we will explore the dynamics of cooperation and cheating in biological systems, and how game theory provides insights into understanding and analyzing these behaviors.

Cooperative Behaviors in Biological Systems

Cooperative behaviors in biological systems involve individuals acting in a way that benefits not only themselves but also others in their social group. Common examples of cooperation can be seen in social insects like bees and ants, where workers collaborate to build and maintain nests, gather food, and care for the brood. Additionally, cooperation is prevalent among animals that engage in reciprocal altruism, such as mutual grooming in primates or reciprocal feeding in vampire bats.

The evolution of cooperation has been a long-standing puzzle in biology, as it raises the "problem of altruism": how can natural selection favor traits that benefit others at a cost to oneself? One explanation lies in the concept of kin selection, where individuals are more likely to cooperate with close relatives who share genetic material. This can be seen in social insects, where workers are often sisters, and their cooperation enhances the survival and reproduction of their shared genes.

Another mechanism that promotes cooperation is direct reciprocity, where individuals engage in mutually beneficial interactions with the expectation of future reciprocation. This is observed in many animal species, where individuals form long-term relationships and engage in behaviors that provide immediate benefits to others, with the understanding that they will receive similar benefits in return.

Cheating and the Problem of Cheaters

While cooperation can be beneficial to a group as a whole, it also presents opportunities for individuals to cheat and exploit the efforts of others without contributing themselves. Cheaters can undermine the stability and viability of cooperative systems, as they reap the benefits of cooperation without incurring the costs. If cheating becomes prevalent, it can lead to the breakdown of cooperation altogether.

An illustrative example of cheating in biological systems is observed in the relationship between cleaner fish and their clients. Cleaner fish form mutualistic interactions with other fish species by removing parasites from their bodies. However, some cleaner fish have been observed to cheat by consuming the mucus of their clients without providing any cleaning services in return. This behavior reduces the benefits received by the client fish and disrupts the cooperative equilibrium.

Game Theory and the Dynamics of Cooperation and Cheating

Game theory provides a powerful framework for understanding the dynamics of cooperation and cheating in biological systems. In a game-theoretic context, individuals are viewed as players who make strategic decisions to maximize their own fitness or reproductive success. The outcomes of these decisions are influenced by the strategies and actions of other players in the game.

One of the most well-known game-theoretic models for studying cooperation and cheating is the Prisoner's Dilemma. In this game, two individuals face the choice of either cooperating or defecting. If both players cooperate, they receive a moderate reward. However, if one player defects while the other cooperates, the defector gains the highest reward while the cooperator suffers the most severe punishment. If both players defect, they receive a lesser reward than if they had both cooperated.

The Prisoner's Dilemma captures the tension between cooperative and selfish behavior. It demonstrates how rational individuals may be tempted to defect and exploit the cooperation of others, even though cooperation would yield the highest collective benefit.

Strategies for Promoting Cooperation and Deterring Cheating

To ensure the stability and sustainability of cooperative systems, mechanisms have evolved to promote cooperation and deter cheating. These mechanisms can be categorized into two main types: direct enforcement and indirect enforcement.

Direct enforcement relies on punishment or negative interactions to deter cheaters. This can be seen in the behavior of cleaner wrasses, a type of cleaner fish. When a client fish detects cheating by a cleaner wrasse, it sometimes punishes the cheater by chasing, biting, or even refusing to interact with it in the future. Such punishment acts as a deterrent and reduces the frequency of cheating in the population.

Indirect enforcement involves mechanisms that incentivize cooperation or provide benefits to cooperators. One such mechanism is reputation, where individuals build a social image based on their past interactions. In certain bird species, for example, individuals that have been observed to help others in the past are more likely to receive help from others in the future. This incentivizes cooperative behavior and reduces the chances of cheating.

Evolutionary Strategies for Dealing with Cheaters

Evolutionary game theory provides insights into the evolutionary strategies that can emerge to address the problem of cheaters in biological systems. One such strategy is conditional cooperation, where individuals are more likely to cooperate in the presence of previous cooperative interactions or when surrounded by cooperators. This allows cooperative behaviors to persist in the presence of occasional cheaters.

Another strategy is the evolution of punishment, where individuals engage in costly behaviors to penalize cheaters. In a game-theoretic context, this can be modeled as a "Punisher vs. Cheater" game, where the punishers incur a cost to punish the cheaters but gain a long-term benefit by deterring future cheating. The presence of punishers in a population can stabilize cooperation by reducing the payoff of cheating and promoting cooperation.

Beyond Biological Systems: Applications of Cooperation and Cheating

The dynamics of cooperation and cheating are not limited to biological systems. They also have implications in various domains, such as human societies, economics, and politics. By studying the principles and mechanisms underlying cooperation and cheating in biological systems, we can gain insights into understanding and solving real-world problems related to collaboration, trust, and the management of public goods.

For example, understanding the conditions that promote cooperation and deter cheating can inform the design of institutions and policies to address collective action problems. Game theory can be applied to study cooperation and cheating in economic transactions, international relations, and social networks, providing strategies for promoting cooperation and mitigating the risks of exploitation.

In conclusion, the study of cooperation and cheating in biological systems offers valuable insights into the dynamics of social behavior and the evolution of stable cooperative systems. Game theory provides a powerful framework for analyzing and understanding these behaviors, elucidating the strategies and mechanisms that promote cooperation and deter cheating. By expanding our understanding of cooperation and cheating in biological systems, we can glean important lessons applicable to a wide range of disciplines and real-world scenarios.

Game Theory and the Evolution of Animal Behavior

In the fascinating world of animal behavior, game theory provides valuable insights into the strategies and interactions that shape the survival and reproduction of different species. By modeling these interactions as games, scientists can understand how animals make decisions, adapt to their environments, and ultimately evolve.

Darwinian Evolution and Game Theory

Before we delve into game theory's role in understanding animal behavior, let's take a moment to appreciate the backdrop against which these interactions occur: Darwinian evolution. Evolutionary biologists acknowledge that the primary goal of any organism is to pass on its genes to the next generation. This drive for reproductive success shapes an animal's behavior, often leading to strategic decision-making.

Game theory provides a framework to understand these strategic interactions, where animals can adopt different strategies and compete for resources, mates, and territory. By studying these interactions, we gain deeper insights into the selection pressures that shape the behavior of various species.

Fitness and Replicator Dynamics

Fitness is a critical concept in evolutionary biology, measuring an organism's reproductive success in a given environment. In the context of game theory, fitness can be seen as the payoffs an individual receives based on its strategies and the strategies of others. The more successful an organism is in passing on its genes, the higher its fitness.

Replicator dynamics, a mathematical model used in evolutionary game theory, helps us understand how different strategies evolve within a population over time. It predicts that strategies with higher payoffs will become more prevalent in a population, while strategies with lower payoffs will diminish.

Evolutionary Stable Strategies: Surviving in a Changing World

In the complex world of animal interactions, certain strategies can become evolutionarily stable, meaning they resist invasion by alternative strategies. These strategies, known as Evolutionary Stable

Strategies (ESS), provide a stable equilibrium point where no individual can gain an advantage by deviating from the prevailing strategy.

For example, in a population of male birds competing for mates, some may exhibit aggressive behavior, while others display more passive strategies. If the aggressive strategy confers a higher mating success and the passive strategy avoids costly fights, a stable equilibrium may be reached where both strategies coexist over time.

The Prisoner's Dilemma in Evolutionary Biology

One of the most famous games used to study animal behavior is the Prisoner's Dilemma. In this game, two individuals must decide whether to cooperate or defect. The payoffs are structured in a way that cooperation would benefit both players, but defection provides a higher payoff if the other player cooperates.

In the context of animal behavior, the Prisoner's Dilemma helps us understand scenarios where cooperation may arise due to repeated interactions. For example, in vampire bats, individuals may share blood meals with others, even if they were unsuccessful in finding their own meal. This reciprocal behavior ensures future cooperation and promotes the survival of the entire group.

Cooperation and Cheating in Biological Systems

Cooperation in nature often raises the question of cheaters—individuals that exploit the cooperative behavior of others without contributing themselves. These cheaters can enjoy the benefits of cooperation without paying the costs, undermining the stability of cooperative strategies.

To address this challenge, scientists have explored mechanisms to deter cheating. One such mechanism is direct reciprocity, where individuals can remember past interactions and adjust their behavior accordingly. By punishing cheaters or rewarding cooperation, the prevalence of cheating can be reduced, maintaining the stability of cooperative strategies.

Game Theory and the Evolution of Animal Behavior

Game theory provides an invaluable tool to study the evolution of animal behavior. It allows us to model and analyze the strategic interactions between individuals, uncovering the underlying dynamics that shape their decision-making.

From the perspective of animal behavior, game theory helps us understand why certain strategies prevail over others, how cooperation evolves, and why conflicts may arise. By applying game theory to real-world examples, such as the sharing of resources, parental care, or territorial disputes, we can gain deeper insights into the complexities of animal behavior and its evolutionary origins.

Real-World Applications of Game Theory in Ecology and Sociology

The study of game theory has far-reaching implications, extending beyond animal behavior to various fields such as ecology and sociology. In ecology, game theory can shed light on predator-prey interactions, optimal foraging strategies, and the evolution of mating systems. In sociology, it helps us decipher the dynamics of cooperation, social norms, and conflict resolution.

Understanding the principles of game theory equips researchers with a powerful toolkit to explore and unravel the mysteries of the natural world. By uncovering the strategic interactions that govern animal behavior, we gain a deeper appreciation for the intricate web of life and our place within it.

In Case You're Curious:

If you're itching to explore more about the intersection of game theory and animal behavior, check out the groundbreaking work of John Maynard Smith and Robert Axelrod. Their research dives into the depths of game theory's application in understanding the evolution of cooperation and the strategic interactions that shape animal societies.

Applications of Evolutionary Game Theory in Ecology and Sociology

Evolutionary game theory is a powerful framework that allows us to understand the dynamics of strategic interactions in various fields, including ecology and sociology. By applying evolutionary principles to game theory, we can gain insights into how different strategies evolve and persist in populations over time. In this section, we will explore the applications of evolutionary game theory in ecology and sociology, shedding light on the fascinating interplay between individuals and their environments.

Evolutionary Game Theory in Ecology

In ecology, individuals often engage in strategic interactions to maximize their fitness and adapt to their ever-changing environments. Evolutionary game theory provides a valuable tool for studying these interactions and understanding the emergence and persistence of different strategies. Let's explore some key applications of evolutionary game theory in ecology.

1. Evolution of Altruism   One of the most intriguing phenomena in ecology is the evolution of altruism, where individuals perform actions that benefit others at a cost to themselves. How can altruistic behaviors persist in a population where self-interest seems to be the prevailing strategy? Evolutionary game theory can shed light on this puzzle.

Consider the classic example of "helper" birds in a population. These birds forego their own reproductive success to assist others in raising their offspring. This behavior seems contradictory from an individualistic perspective, as it reduces the helper's own fitness. However, when we analyze this situation using game theory, we find that altruistic behavior can evolve under certain circumstances.

By modeling the interactions between helpers and breeders, researchers have shown that altruistic behavior can be favored when the cost-to-benefit ratio is within a certain range. When the cost of helping is relatively low compared to the benefit gained by the recipients, natural selection can favor the spread of altruistic genes. This finding demonstrates how evolutionary game theory can provide valuable insights into the evolution of altruism in ecological systems.

2. Predator-Prey Dynamics   Predator-prey interactions play a crucial role in shaping ecosystems. The dynamics between predators and their prey can be modeled using evolutionary game theory to understand how different strategies evolve in response to each other.

For instance, consider a population of hares and lynxes. Hares exhibit anti-predator behaviors, such as increased vigilance and defensive hiding, while lynxes employ hunting strategies to maximize their feeding success.   By analyzing the payoff matrix of this predator-prey interaction, we can determine the evolutionarily stable strategies for both species.

Through the application of evolutionary game theory, we can uncover the conditions under which populations of predators and prey can coexist and reach a stable equilibrium. This understanding is crucial for predicting the consequences of environmental changes and developing effective conservation strategies.

3. Evolution of Cooperation in Social Insects Social insects, such as ants and bees, exhibit remarkable levels of cooperation within their colonies. Evolutionary game theory has proven invaluable in unraveling the mechanisms underlying the evolution and maintenance of cooperation in these complex societies.

One example is the decision to become a worker ant rather than a reproductive queen. Worker ants sacrifice their personal reproduction to serve the queen and the colony as a whole. Evolutionary game theory provides a framework for understanding how this stable caste system can emerge and persist.

By modeling the interactions among different castes and quantifying the costs and benefits of each role, researchers have demonstrated how the evolution of worker behavior can be explained by kin selection and indirect fitness benefits. This exploration highlights the power of evolutionary game theory in explaining the cooperative behaviors observed in social insects.

Evolutionary Game Theory in Sociology

Societies are full of strategic interactions, from cooperation within groups to conflicts between different factions. Evolutionary game theory offers a valuable lens through which we can analyze and understand various social dynamics. Let's explore the applications of evolutionary game theory in sociology.

1. **Cooperation and Punishment** Cooperation lies at the heart of human societies, but it can be undermined by the temptation to exploit the efforts of others without contributing oneself. How can cooperation be sustained in environments where selfish behavior seems advantageous?

Evolutionary game theory provides insights into this question by examining different mechanisms that promote cooperation. One such mechanism is punishment, where individuals incur costs to penalize defectors. By modeling these interactions as games, researchers have discovered that the introduction of punishment can foster the evolution of cooperation in populations.

For example, consider the public goods game, where individuals contribute resources for the common good. Without any mechanisms to enforce cooperation, free-riders may exploit the contributions of others. However, when punishment is allowed, cooperators can punish defectors, disincentivizing selfish behavior and promoting cooperation. This finding illustrates the power of evolutionary game theory in explaining the role of punishment in sustaining cooperation within societies.

2. **Cultural Evolution and Social Norms** In human societies, cultural evolution plays a significant role in shaping social norms and behaviors. Evolutionary game theory provides a framework for studying how cultural traits propagate, evolve, and influence societal dynamics.

For instance, consider the evolution of fairness norms. Experimental studies employing the Ultimatum Game, where participants divide a sum of money, have shown that people tend to reject unfair offers. This behavior can be understood through evolutionary game theory by considering the long-term consequences of rejecting unfair offers.

By integrating cultural evolution into game theory models, researchers have revealed how fairness norms can emerge and stabilize in populations over time. This interdisciplinary approach provides a deeper understanding of the cognitive mechanisms underlying human decision-making and the emergence of social norms.

3. Social Contagion and Spread of Behaviors The adoption of new behaviors often spreads through social networks, creating a cascading effect within societies. Evolutionary game theory can help us understand the dynamics of behavior adoption and social contagion.

Consider the spread of pro-environmental behaviors, such as recycling or reducing energy consumption. By modeling these behaviors as games, researchers have shown that the adoption of pro-environmental strategies can spread through populations under certain conditions.

The structure of social networks, the influence of neighbors, and the perceived benefits of adopting a behavior all play vital roles in shaping the outcomes. By investigating the dynamics of behavior adoption using evolutionary game theory, policymakers can design interventions to promote the adoption of sustainable behaviors at a larger scale.

Summary

Evolutionary game theory provides a powerful framework for understanding strategic interactions in ecological and sociological systems. By considering the evolutionary pressures acting on individuals, we can gain insights into the emergence, persistence, and evolution of different strategies. In ecology, evolutionary game theory helps us understand phenomena such as the evolution of altruism, predator-prey dynamics, and cooperation in social insects. In sociology, it provides valuable insights into the evolution of cooperation, the role of punishment and cultural norms, and the spread of behaviors through social networks.

These applications of evolutionary game theory highlight the interdisciplinary nature of this field and its potential to enrich our understanding of complex social and ecological systems. By combining insights from ecology, sociology, and evolutionary biology, we can tackle real-world challenges and develop innovative solutions informed by the principles of game theory.

Strategic Interaction and Game Theory

Bargaining and Competitive Strategy

The Ultimatum Game: Fairness and Rationality

In this section, we will explore the concept of fairness in game theory through an intriguing game known as the Ultimatum Game. We will discuss how fairness and rationality interplay in this game, and delve into its real-world applications.

Understanding the Ultimatum Game

The Ultimatum Game is a simple yet thought-provoking game that involves two players - the proposer and the responder. The rules are as follows:

1. The proposer is given a sum of money, say $100, which they can divide between themselves and the responder. 2. The proposer makes an offer, specifying how much they will keep and how much they will offer to the responder. 3. The responder can either accept or reject the offer. 4. If the responder accepts, both players receive the allocated amounts. 5. If the responder rejects, neither player receives anything.

At first glance, you might think that the rational thing for the responder to do is to always accept any offer, no matter how unfair it may seem. After all, receiving something is better than receiving nothing, right? Well, it turns out that human behavior often deviates from strict rationality, and that's where the fascinating dynamics of the Ultimatum Game come into play.

Fairness vs. Rationality

In the Ultimatum Game, fairness takes center stage. The responder's decision to accept or reject an offer depends on their perception of fairness and their desire to punish unfairness. This phenomenon has been observed in numerous experimental studies conducted in various cultural contexts.

Contrary to purely rational behavior, responders tend to reject low offers that they deem unfair. This behavior can be attributed to the desire for fairness and the aversion to inequality. Responders may reject an offer, even though it means receiving nothing, to punish the proposer's perceived unfairness.

This deviation from strict rationality challenges the traditional notion of economic rationality, which assumes individuals always act in their own self-interest. The Ultimatum Game reveals that considerations of fairness and social norms can play an influential role in decision-making.

Real-World Applications

The Ultimatum Game has important implications for various real-world scenarios, shedding light on how fairness and rationality interact in different contexts. Let's explore some examples:

1. Salary Negotiations: During salary negotiations, an employer may offer a low raise or bonus to an employee. If the employee perceives this offer as unfair, they might reject it, even if it means forgoing the additional income. This can be seen as a form of punishment for the perceived unfairness.

2. Sharing Resources: In situations where resources need to be divided, such as in the distribution of government aid or communal goods, the fairness of the allocation can significantly impact people's reactions. If individuals perceive the distribution as unfair, they may refuse to participate or actively protest against it.

3. Social Contracts: The Ultimatum Game provides insights into social contract theory. It highlights the importance of fairness in establishing equitable social norms and expectations. Individuals are more likely to cooperate and comply with a system they perceive as fair, rather than one that is seen as unjust or exploitative.

4. Negotiations and Bargaining: The dynamics of the Ultimatum Game can be applied to negotiating situations, such as business deals or

international diplomacy. The perceived fairness of offers plays a crucial role in the success or failure of negotiations. Understanding the role of fairness can help negotiators craft more acceptable and mutually beneficial agreements.

Tricky Situations and Resolutions

While the Ultimatum Game provides valuable insights into fairness and rationality, it also poses some interesting challenges. Consider the following scenarios:

1. Incomplete Information: What if the proposer is unaware of the responder's fairness preferences? In such cases, the proposer might make an initial offer that doesn't align with the responder's expectations, leading to a rejection. Open communication and understanding each other's preferences can help resolve this situation.

2. Cultural Variations: Different cultures may have varying concepts of fairness and equity. Cultural norms and values can influence people's decision-making in the Ultimatum Game. Researchers have found that individuals from certain cultures, such as Western cultures, tend to reject low offers more frequently than individuals from other cultures. Recognizing cultural diversity is crucial in interpreting the outcomes of the game.

3. Strategic Behavior: Players may engage in strategic behavior, attempting to manipulate the perception of fairness to their advantage. For example, a proposer might make an initially low offer, expecting the responder to reject it. This rejection could push the proposer to make a more favorable offer in subsequent rounds, enabling them to strike a better deal overall.

Further Reading and Resources

If you want to dive deeper into the Ultimatum Game and its implications, the following resources can provide you with additional insights:

1. Book: "Fairness and Evolutionary Psychology" by Peter K. Smith and Anna Machin explores the role of fairness in human evolution and its impact on decision-making.

2. Research Paper: "Ultimatum Bargaining: A Survey and Comparison of Experimental Results" by Werner Güth and Rolf

Schmittberger provides an extensive review of experimental studies on the Ultimatum Game.

3. Online Resource: The TED Talk by Dan Ariely, titled "What makes us feel good about our work," delves into the concept of fairness and its influence on motivation and productivity.

4. Academic Journal: The Journal of Economic Behavior & Organization regularly publishes studies on fairness, social preferences, and decision-making, offering a wealth of research on these topics.

Remember, the Ultimatum Game highlights the complexity of human behavior in decision-making, reminding us that rationality alone cannot fully explain our choices. Fairness and social norms shape our interactions, creating a rich and dynamic landscape for game theory exploration.

Auctions: Bidding for Success

In the world of commerce, auctions have long been a popular method for buying and selling goods, where potential buyers compete to obtain the items up for sale. While auctions may seem like a simple process of placing bids, winning, and paying, the underlying dynamics can be much more intricate. This is where game theory comes into play, helping us understand the strategies and tactics that can lead to success in auctions.

Understanding Auctions

An auction is essentially a competitive process where individuals or organizations bid for a specific item or service. The auctioneer sets the rules and conducts the auction, while participants place their bids with the goal of acquiring the item at the lowest possible price. There are different types of auctions, each with its own set of rules and objectives.

One common type of auction is the English auction, also known as an open-outcry auction. In this type of auction, participants openly announce their bids, and the highest bidder wins the item. Another type is the sealed-bid auction, where participants submit their bids in a sealed envelope. The highest bidder, when the envelopes are opened, wins the item.

Auction Formats and Strategies

Auction theory explores the optimal bidding strategies participants can employ to maximize their chances of success. The most well-known and studied auction format is the Vickrey auction, also called a second-price sealed-bid auction. In this auction, the highest bidder wins the item but pays the price of the second-highest bid. This format encourages bidders to truthfully reveal their valuations, as bidding higher than their true value does not increase their chances of winning.

The optimal strategy in a Vickrey auction is to bid your true valuation of the item. It ensures that you pay a price equal to or slightly lower than your actual valuation, maximizing your potential gain. However, strategic considerations come into play when there is uncertainty regarding the valuations of other bidders.

Strategic Considerations and Bid Manipulation

Game theory teaches us that understanding the motivations and strategies of other bidders can give you an advantage in auctions. In situations where valuations are uncertain, it may be beneficial to manipulate your bids strategically. For example, bidding slightly higher than your true valuation can serve to discourage other bidders, making them believe that you value the item more than they do.

Bid shading is another strategy commonly employed in auctions. It involves deliberately underbidding your true valuation to secure the item at a lower price. However, bid shading can be risky. If other bidders catch on to your strategy, they may bid aggressively, potentially resulting in losing the item altogether.

Common Auction Pitfalls

While auctions offer opportunities for obtaining goods or services at favorable prices, there are some pitfalls to be aware of. One common trap is known as the winner's curse, which occurs when the winner of an auction overpays due to an overly optimistic estimation of the item's value. To avoid falling into this trap, it's crucial to conduct thorough research and have a realistic understanding of the item's worth.

Another challenge is bid collusion, where participants conspire to keep prices artificially low. This practice undermines the competitive nature of auctions and is often illegal. Auction regulators actively monitor and investigate suspicious bidding activities to maintain a fair and transparent marketplace.

Real-World Examples

Auctions are not just a theoretical concept — they play a significant role in various industries. For instance, in art auctions, bid manipulation strategies are often employed by experienced bidders to create an atmosphere of competitiveness and inflate prices.

Government auctions are another area where game theory and auctions intersect. For example, when allocating radio frequencies or licenses, governments use auction mechanisms to ensure fair distribution while maximizing revenue. The spectrum auction held by the Federal Communications Commission (FCC) in the United States is a well-known example.

Conclusion

Auctions are dynamic environments where strategic decision-making can greatly impact outcomes. Game theory provides a framework for understanding and analyzing bidding strategies in various auction formats. By studying the principles, pitfalls, and real-world applications of auctions, individuals can enhance their chances of bidding for success. Remember, in auctions, being smart and strategic is the key to coming out on top.

Oligopoly: Competition and Collusion

In the world of economics, competition is the name of the game. And when it comes to market structures, one of the most interesting and complex ones is oligopoly. So, get ready to dive into the fascinating world of oligopoly, where competition and collusion collide!

Understanding Oligopoly

Oligopoly refers to a market structure dominated by a small number of firms. Unlike a monopoly, where there is only one dominant player, and perfect competition, where there are many small players, oligopoly strikes a balance between the two extremes.

In an oligopoly, a few powerful firms have a significant market share and can influence the market conditions by their actions. These firms often produce differentiated products or offer unique services, giving them some degree of market power.

Competition in Oligopoly

Competition in oligopoly is intense and fierce. Each firm in an oligopolistic market must consider the actions and reactions of its competitors when making decisions. This interdependence among firms sets the stage for strategic behavior and game theory analysis.

Game theory, which we explored in earlier chapters, plays a crucial role in understanding and analyzing the behavior of firms in an oligopoly. It helps us analyze the strategic interactions between firms and determine the most rational choices they can make.

Game Theory in Oligopoly

In the game of oligopoly, firms engage in various strategic moves, such as pricing decisions, production levels, advertising, product differentiation, and market entry. Each decision can affect the profits of all the firms operating in the market.

To capture the interplay between firms, game theory provides us with a powerful tool called the "Prisoner's Dilemma." This concept helps us understand the incentives and potential outcomes when firms must decide between competing aggressively or cooperating.

In the Prisoner's Dilemma, each firm faces a choice: to collude or to compete. If all the firms choose to compete, they may engage in a "price war," leading to lower profits for all. On the other hand, if they collude and agree to fix prices or restrict output, they can collectively earn higher profits.

Collusion in Oligopoly

Collusion refers to an agreement or understanding among competing firms to limit competition and act cooperatively. In an oligopolistic market, firms may collude through explicit agreements, such as cartels, or tacit understandings, where cooperation happens without formal agreements.

The goal of collusion is to increase profits by reducing competition. By coordinating their actions, firms can collectively raise prices, restrict output, allocate market shares, or engage in other strategies that benefit all involved.

However, collusion is not always easy to achieve or sustain. Each firm has an incentive to cheat and undercut the others to gain a competitive advantage. This "prisoner's dilemma" creates a delicate balance between cooperation and self-interest.

The Price-Fixing Game

To better understand collusion in oligopoly, let's explore a classic example: the price-fixing game. Imagine two firms, Alpha and Beta, operating in an oligopolistic market. Both firms have two options: set a high price (H) or set a low price (L).

If both firms choose to set a high price, they both earn moderate profits. If both choose to set a low price, they both earn low profits due to intense competition. However, if one firm sets a high price while the other sets a low price, the low-price firm earns high profits, while the high-price firm suffers.

This scenario creates a dilemma for both firms. If they both cooperate and set high prices, they can maximize their combined profits. However, each firm has an incentive to deviate from the agreement and set a lower price, earning higher profits individually while hurting the other firm's bottom line.

Detecting Collusive Behavior

Detecting collusive behavior in the real world can be challenging. Firms engaged in collusion often try to hide their actions, making it difficult for regulators to uncover illegal agreements. However, there are some signs that may indicate collusive behavior:

1. Stable prices: If prices remain stable despite fluctuations in demand or production costs, it could be a sign of collusion.

2. Uniform pricing: If competing firms charge identical or very similar prices, collusion might be at play.

3. Restricted output: When firms limit production or allocate market shares without clear reasons based on cost or efficiency, collusion is a possible explanation.

4. Parallel behavior: If firms mimic each other's actions, such as price changes or advertising campaigns, collusion becomes more likely.

Regulating Oligopoly

Due to the potential negative effects of collusion on consumer welfare, many jurisdictions have regulations to prevent and punish anticompetitive behavior in oligopoly markets. Regulatory bodies, such as competition commissions, monitor and investigate firms suspected of collusive practices.

Penalties for collusion can be severe, including hefty fines, disbanding cartels, or even criminal charges. By deterring collusion, regulators aim to promote fair competition and protect consumer interests.

Real-World Examples

Collusion in oligopoly has been documented in various industries throughout history. Here are a few notable examples:

1. OPEC (Organization of the Petroleum Exporting Countries): OPEC is an international cartel consisting of oil-producing countries that coordinate production levels to influence oil prices.

2. LCD Panel Industry: Several major manufacturers were found guilty of price-fixing in the LCD panel industry, leading to billions of dollars in fines.

3. Airline Industry: There have been accusations of collusion among airlines, involving practices like price signaling and capacity coordination.

These examples highlight the challenges faced by regulators in detecting and preventing collusive behavior in oligopoly markets.

Exercise

Consider a hypothetical oligopolistic market with three firms: Alpha, Beta, and Gamma. Assume they can choose two strategies: produce a large quantity (L) or produce a small quantity (S). The payoff matrix below shows the profits each firm earns based on their choices:

	Alpha	Beta	Gamma
Alpha	12, 12	10, 15	15, 10
Beta	15, 10	14, 14	8, 12
Gamma	10, 15	12, 8	13, 13

1. Determine the Nash equilibrium in this game. What strategies should each firm choose?

2. Is there a dominant strategy for any of the firms? Explain your answer.

3. Suppose the firms decide to collude and agree to produce a small quantity. Will this agreement benefit all firms? Why or why not?

Conclusion

Oligopoly is a unique market structure that challenges traditional notions of perfect competition and monopoly. In this fascinating world, the competition is fierce, and the decisions made by each firm have a profound impact on the market as a whole.

Understanding the strategic interactions and dynamics of oligopoly requires the application of game theory, particularly the concepts of collusion and competition. By delving into the intricacies of these concepts, we can navigate the complexities of oligopoly and gain insights into real-world economic phenomena.

So, buckle up and get ready to dive into the strategic world of oligopoly, where competition and collusion dance a delicate tango of self-interest and cooperation.

Incentives and Agency Theory in Strategic Interaction

In the world of strategic interaction, where individuals or entities make decisions that affect not only their own outcomes but also the outcomes of others, incentives play a crucial role. Incentives can be thought of as the motivating factors that drive individuals to act in a certain way. They can be positive, like rewards or benefits, or negative, like punishments or costs. In this section, we will explore the concept of incentives and how they relate to agency theory in the context of strategic interaction.

Understanding Incentives

Incentives are an essential aspect of strategic interaction because they shape the behavior of individuals or entities involved in a game. They can influence players' decisions by altering their perceived payoffs, thereby affecting their strategies. In the realm of game theory, incentives can be categorized into two types: explicit and implicit incentives.

Explicit incentives are rewards or punishments that are explicitly stated or contractually agreed upon. These incentives are typically monetary or material in nature. For example, a company might offer a performance bonus to motivate its employees or impose fines for non-compliance with company policies. In explicit incentive systems, the correlation between desired behavior and rewards or punishments is clearly defined.

Implicit incentives, on the other hand, are more subtle and rely on social norms, reputation, or other non-monetary factors to motivate behavior. These incentives are often based on trust, reciprocity, and the desire to maintain social relationships. For instance, in a group project, team members may be motivated to contribute their fair share out of a sense of responsibility and in order to maintain a positive reputation among their peers.

Principal-Agent Problem

One of the key concepts related to incentives in strategic interaction is the principal-agent problem. This problem arises when one party, the principal, enlists the help of another party, the agent, to carry out a task on their behalf. The challenge lies in aligning the interests of the principal and the agent, as their objectives may not necessarily be perfectly aligned. The principal seeks to maximize their own utility, while the agent may have their own goals and motivations.

In the context of agency theory, the principal-agent problem is often examined in the context of a contractual relationship, where the principal delegates decision-making authority to the agent. The principal relies on incentives to align the agent's behavior with their own interests. However, there is always a risk that the agent may act opportunistically, pursuing their own self-interest at the expense of the principal.

Solutions to the Principal-Agent Problem

To mitigate the principal-agent problem and align the interests of the principal and the agent, a variety of mechanisms and strategies have been developed. These include:

1. **Monitoring and Reporting**: The principal can employ monitoring mechanisms to keep track of the agent's activities and ensure compliance with the agreed-upon objectives. This can involve regular reporting, performance evaluations, or the use of technology to monitor behavior.

2. **Incentive Alignment**: Designing incentive systems that align the goals of the agent with those of the principal is crucial. By offering rewards or punishments that are directly tied to desired outcomes, the principal can motivate the agent to act in ways that maximize their joint interests.

3. **Contractual Arrangements**: Well-designed contracts can help mitigate the principal-agent problem by explicitly outlining the roles, responsibilities, and expected behavior of both parties. Clear and enforceable contracts reduce uncertainty and provide a framework for resolving conflicts of interest.

4. **Reputation and Trust**: Building trust and maintaining a good reputation can provide implicit incentives for agents to act in the best interests of the principal. A history of fair and trustworthy behavior can serve as a powerful motivator for agents to maintain their reputation and foster long-term relationships.

5. **Intra-Firm Communication and Collaboration**: Effective communication and collaboration within an organization can help align the interests of principals and agents. By fostering a culture of transparency, information sharing, and teamwork, organizations can enhance cooperation and reduce conflicts of interest.

Real-World Examples

To illustrate the concepts of incentives and agency theory in strategic interaction, let's consider a few real-world examples.

In the financial industry, investment managers often act as agents for their clients (the principals). These managers are entrusted with making investment decisions on behalf of their clients, with the goal of maximizing returns. To align the interests of the managers and the clients, performance-based incentive structures, such as charging a percentage of the investment gains or offering performance bonuses, are commonly used. This helps to create an alignment between the interests of the agent and the principal.

In the field of healthcare, the principal-agent problem arises between patients (the principals) and healthcare providers (the agents). Patients rely on doctors to make decisions that are in their best interest, taking into account their medical needs and preferences. However, doctors may also have their own incentives, such as financial interests or pressure from pharmaceutical companies. To address this problem, regulatory bodies have implemented measures to ensure that doctors prioritize patient welfare over other considerations. These measures include enforcing ethical guidelines, imposing penalties for unethical behavior, and promoting transparency in doctor-patient relationships.

Conclusion

Incentives have a significant impact on the behavior of individuals or entities involved in strategic interaction. To ensure alignment between interests and promote cooperation, it is crucial to design effective incentive systems and address the principal-agent problem. By understanding the underlying principles of agency theory and leveraging appropriate strategies, individuals and organizations can navigate the complex landscape of strategic interaction with greater success.

Signaling and Screening Games

In the world of game theory, signaling and screening games are an intriguing subset of strategic interactions. These games involve scenarios in which one party possesses private information that they can use strategically to influence the decisions of other players. Signaling games occur when an informed player sends a signal to convey their private information to an uninformed player, while screening games involve the inverse situation, where the uninformed player designs a screening mechanism to extract information from the informed player.

Principles of Signaling and Screening Games

In signaling games, the informed player aims to reveal their type or quality through a signal, which can be either costly or costless. The signal acts as a credible indicator, allowing the uninformed player to make an optimal decision based on this information.

On the other hand, in screening games, the uninformed player devises a screening mechanism to separate the informed player's different types or qualities. By designing a screening strategy that induces the informed player to reveal their private information truthfully, the uninformed player can make a more informed decision.

Signaling and screening games are often associated with asymmetric information, where one party possesses information that the other party does not have. These types of games are prevalent in various fields, including economics, finance, marketing, and even in interpersonal relationships.

The Lewis Signaling Model

One well-known example of signaling games is the Lewis Signaling Model. This model illustrates the dynamics of signaling between a job seeker and a potential employer. The job seeker possesses private information about their ability and skills, while the employer relies on observable characteristics during the hiring process.

In this model, the job seeker aims to convey their quality to the employer through a costly signal, such as obtaining a higher education degree. The signal acts as a credible indicator of the job seeker's ability, allowing the employer to make an informed hiring decision.

However, the challenge lies in distinguishing between individuals who have a genuine high ability and those who are simply mimicking the signal. This informational asymmetry creates a strategic interaction between the job seeker and the employer.

The Principal-Agent Problem

Another important application of signaling and screening games is in the context of the principal-agent problem. In this scenario, a principal (such as a company owner) delegates a task to an agent (such as an employee), who has private information about their effort level or capabilities.

The principal wants the agent to perform the task diligently or exert high effort. However, the principal cannot directly observe the agent's effort, leading to a problem of adverse selection or moral hazard.

To address this problem, the principal can design a screening mechanism, such as offering different contract types or incentives, to induce the agent to reveal their private information. By properly aligning the incentives, the principal can encourage the agent to act in the desired way.

Real-Life Examples of Signaling and Screening Games

Signaling and screening games can be found in various real-life situations. One example is college admissions, where students attempt to signal their abilities and qualities through their academic achievements, extracurricular activities, and recommendation letters. Admissions officers, in turn, screen the applicants using these signals to select students who are likely to succeed at the university.

In the context of job interviews, candidates often aim to signal their skills and competence to potential employers through their resumes, cover letters, and interview performance. Employers, on the other hand, use screening techniques such as skill tests, background checks, and reference checks to gather additional information about the candidates.

In the world of finance, lenders use screening techniques to evaluate borrowers' creditworthiness and risk profiles. They gather information through credit scores, income statements, and collateral assessments to make lending decisions.

Tricks and Caveats in Signaling and Screening Games

In signaling and screening games, it is crucial to consider the cost of the signal or screening mechanism. Costly signals are more likely to be credible, as they demonstrate the sender's willingness to pay for their claim. However, a high cost might deter honest players from sending signals, leading to adverse outcomes.

Additionally, players must be mindful of the potential for strategic manipulation and opportunistic behavior. The uninformed player may design a screening strategy that induces false signaling or hides their true private information. This strategic behavior can lead to signaling and screening failures, which in turn can have negative consequences.

Understanding the underlying incentives and strategic considerations is key to navigating the complexities of signaling and screening games. By analyzing the players' incentives and constraints, researchers and decision-makers can design more effective signaling and screening mechanisms.

Exercises

1. Consider the following scenario: A company is hiring for a position that requires a specific skill set. Job applicants can only signal their skill set through their work experience. The company wants to attract qualified candidates. What type of screening mechanism could the company use to separate the high-skill individuals from those with lower skills?

2. Imagine you are a student applying for a scholarship. How can you strategically signal your potential and qualifications to the scholarship committee? What factors should you consider when designing your signaling strategy?

3. In the context of online dating, individuals often create dating profiles to attract potential partners. How can these profiles be seen as signals or screening mechanisms? What are some potential pitfalls or vulnerabilities in this process?

Resources

1. "Signaling and Screening in Competitive Markets" by Eric S. Maskin 2. "Game Theory" by Drew Fudenberg and Jean Tirole 3. "Information and Learning in Markets" by Xavier Vives 4. "Economics of Strategy" by David Besanko and David Dranove

By understanding the principles of signaling and screening games, we can gain insights into decision-making processes in various domains. Recognizing the strategic interactions at play and the importance of credible information can help us navigate real-life scenarios where private information and asymmetric knowledge are at play. So, the next time you encounter a situation where signals and screening mechanisms are at play, remember to stay sharp and play strategically!

Advertising and Branding Strategies

Advertising and branding play a crucial role in the business world, helping companies differentiate themselves from their competitors and attract customers. In this section, we will explore the principles of advertising and branding strategies and how they can be analyzed using game theory.

The Importance of Advertising

Advertising serves as a powerful tool for companies to communicate with their target audience. By creating persuasive messages, companies can influence consumer behavior, build brand awareness, and increase market share. However, developing effective advertising strategies requires a deep understanding of consumer preferences, competition, and market dynamics.

Game Theory in Advertising

Game theory provides a framework for analyzing strategic interactions between different players, and it can also be applied to advertising and branding strategies. In the context of advertising, companies can be seen as players who strategically choose their actions to maximize their expected outcomes.

Strategic Decision-Making in Advertising

When developing advertising strategies, companies must consider various factors, such as the target audience, budget constraints, and competitive landscape. Game theory can help companies make strategic decisions by considering the actions and reactions of their competitors.

Differentiating through Branding

Branding is another important aspect of advertising strategy. It involves creating a unique image or identity for a product or a company that sets them apart from others in the market. Effective branding can enhance brand recognition, loyalty, and perceived value.

Advertising Campaigns: Choosing the Right Message

Creating a compelling advertising campaign involves crafting the right message that resonates with the target audience. Game theory can be used to analyze the impact of different advertising messages on consumer behavior and the competition's response.

Advertising Auctions and Bidding Strategies

In online advertising, companies often bid for ad placements on various platforms. Game theory can help companies determine the optimal bidding strategies in these auctions, taking into account factors such as bid prices, quality scores, and expected click-through rates.

Cooperative Advertising and Partnerships

Cooperative advertising involves collaborating with other companies to share the costs and benefits of advertising campaigns. Game theory can be used to model these cooperative interactions, analyzing the optimal allocation of resources and the division of advertising efforts.

Ethical Considerations and Consumer Perception

When developing advertising and branding strategies, companies must also consider ethical considerations and consumer perception. Misleading or manipulative advertising practices can harm a company's reputation and lead to legal consequences. Game theory can provide insights into ethical decision-making in the advertising industry.

Real-World Examples

To illustrate the concepts discussed in this section, let's consider a real-world example. Imagine two competing beverage companies, A and B, launching advertising campaigns for their new soft drinks. Company A decides to invest heavily in a Super Bowl ad, aiming to generate maximum brand exposure. In response, Company B strategically chooses to sponsor a popular music festival, targeting a different segment of the market. By analyzing the strategic decisions of both companies using game theory, we can gain insights into their potential outcomes and competitive dynamics.

Summary

In this section, we explored the role of advertising and branding strategies in the business world. We discussed how game theory can be applied to analyze the strategic interactions between companies in the advertising industry. By understanding the principles of advertising and branding strategies, companies can make informed decisions that maximize their advertising effectiveness and ultimately drive business success.

Game Theory in Marketing and Consumer Behavior

In the world of marketing and consumer behavior, game theory offers a fresh and insightful perspective. By applying the principles of game theory, marketers can better understand the strategic interactions between consumers and companies. This section explores the various ways in which game theory can shed light on marketing strategies, consumer decision-making, and market outcomes.

Strategies and Tactics in Marketing

Every marketing campaign requires careful planning and execution to achieve desired outcomes. Game theory provides a framework for analyzing the strategies and tactics employed by marketers in a competitive marketplace.

One key concept in game theory is the notion of a dominant strategy. A dominant strategy is a course of action that yields the highest payoffs for a player regardless of the strategies chosen by other players. In the context of marketing, a dominant marketing strategy is one that consistently outperforms its competitors.

For example, consider a scenario where two smartphone companies, Apple and Samsung, are competing for market share. Apple launches a new iPhone model with innovative features, while Samsung introduces a similar model at a lower price point. Both companies aim to maximize their market share. By employing game theory, marketers can analyze the strategic interactions between Apple and Samsung to determine the dominant marketing strategy.

Pricing and Competition

Price competition is a crucial aspect of marketing strategy, and game theory offers valuable insights into pricing decisions. In a competitive market, companies face the challenge of setting prices that attract customers while maximizing profits.

One common pricing scenario is that of a duopoly, where two companies dominate the market. Game theory provides tools to analyze the strategic interactions between these companies, considering factors such as pricing, product differentiation, and consumer behavior.

One classic example in marketing is the prisoner's dilemma of pricing. In this dilemma, two companies have the option to either set high prices and maximize profits individually or engage in a price war and earn lower profits. The dilemma arises because if one company chooses to lower prices, it gains a competitive advantage, but both companies may end up with lower profits. Game theory helps marketers understand the incentives and potential outcomes of different pricing strategies.

Advertising and Branding Strategies

Game theory also plays a crucial role in understanding advertising and branding strategies. Marketers aim to create brand loyalty, increase market share, and attract new customers through effective advertising campaigns. Game theory offers insights into the strategic interactions that occur between companies' advertising efforts in a competitive market.

In game theory, there is a concept called an advertising game, which models the competition between companies in terms of their advertising decisions. Companies must decide how much to spend on advertising and how to craft their advertising messages to gain a competitive advantage.

For example, consider a scenario where two soft drink companies, Coca-Cola and Pepsi, are vying for market dominance. They must determine how much to invest in advertising, knowing that their competitor will also be investing in similarly aggressive advertising campaigns. Game theory helps marketers analyze these interactions and identify optimal advertising strategies.

Consumer Decision-Making

Understanding consumer decision-making is a fundamental aspect of marketing. Game theory provides a framework for analyzing the choices consumers make and the factors that influence their decision-making process.

Game theory highlights the importance of considering both individual decision-making and the interaction between consumers. It allows marketers to model consumer behavior in various scenarios, such as product launches, promotions, and pricing strategies. By understanding consumer decision-making, marketers can tailor their strategies to appeal to specific consumer preferences.

Market Outcomes

Game theory also sheds light on market outcomes, including market structures and competitive dynamics. By analyzing strategic interactions between companies, marketers can gain insights into market equilibrium and the likely outcomes of different competitive scenarios.

For instance, game theory can help explain the emergence of oligopolies, where a small number of companies dominate the market. By modeling the strategic interactions between these companies, marketers can better understand the factors that lead to market concentration and its implications for consumer choice.

Additionally, game theory can reveal how market outcomes are influenced by factors such as product differentiation, entry barriers, and consumer preferences. This knowledge allows marketers to anticipate market trends and make informed decisions to gain a competitive advantage.

Real-World Applications

Game theory has found numerous applications in real-world marketing situations. One notable example is the use of dynamic pricing strategies by companies such as Uber and Airbnb. These companies employ algorithms that adjust prices based on demand and supply conditions, taking into account the choices and reactions of both consumers and competitors. Game theory provides the theoretical foundation for understanding and optimizing these dynamic pricing strategies.

Another application is the analysis of brand positioning and market segmentation. Marketers use game theory to identify the optimal positioning of their brands in relation to competitor brands. By understanding consumer preferences and competition dynamics, marketers can strategically position their brands to maximize market share and customer loyalty.

Conclusion

Game theory has become an invaluable tool in the field of marketing and consumer behavior. By analyzing the strategic interactions between consumers and companies, game theory helps marketers develop effective strategies, understand market outcomes, and gain a competitive edge.

Through the application of game theory principles, marketers can strategically navigate the complexities of the marketplace, anticipate the reactions of competitors and consumers, and make informed decisions that drive business success. By embracing game theory, marketers can play a smart game and achieve their desired outcomes in the ever-evolving world of marketing and consumer behavior.

Reputation and Trust in Strategic Interactions

In strategic interactions, reputation and trust play a crucial role in shaping the decisions and outcomes of the players involved. In this section, we will explore the significance of reputation and trust in game theory, their impact on strategic interactions, and how they can be built and maintained.

The Importance of Reputation

Reputation can be thought of as the collective judgment or evaluation that others hold about an individual or entity based on their past actions or behavior. It serves as a crucial signal in decision-making, influencing how others perceive and interact with a player.

In strategic interactions, reputation acts as a form of social capital. It can be a valuable asset that enhances a player's credibility, influence, and opportunities for cooperation. A good reputation can lead to increased trust, cooperation, and better outcomes in games.

Building and Maintaining Reputation

Reputation is not built overnight but rather established and maintained over time through consistent actions and behavior. It requires a player to act in a trustworthy and reliable manner, consistently delivering on their promises, and demonstrating ethical conduct.

One way to build and maintain a positive reputation is through repeated interactions. When players regularly interact with each other, they have the opportunity to observe and learn about each other's behavior, leading to the formation of expectations based on past experiences. These expectations then shape future interactions, as players adjust their strategies based on the reputation of their opponents.

Another mechanism for building reputation is through third-party endorsements or recommendations. When an impartial observer or authority endorses a player's trustworthiness, it can enhance their reputation and increase the likelihood of cooperation. This is commonly seen in online marketplaces or social media platforms, where user ratings and reviews serve as signals of reputation.

Trust as a Strategy

Trust is an essential component of strategic interactions. It refers to the belief or confidence that one player has in the actions, intentions, and reliability of another player. Trust allows players to engage in cooperative behaviors, forming alliances and agreements to achieve mutually beneficial outcomes.

In game theory, trust can be viewed as a strategic choice. Players may choose to trust based on their assessment of the reputation, credibility, and past actions of their opponents. Trusting can be risky, as it entails the possibility of being exploited or betrayed. However, the potential benefits of cooperation and long-term gains often outweigh the risks.

Building and Maintaining Trust

Building trust in strategic interactions requires a delicate balance between risk and reward. Players must accurately assess the trustworthiness of their counterparts while safeguarding their own interests.

One approach to building trust is through repeated interactions. By engaging in a series of games or negotiations, players can gradually establish trust through a history of reliable and cooperative behavior. This is known as tit-for-tat strategy, where players begin by cooperating and then mirror their opponent's previous move in subsequent rounds.

Communication also plays a crucial role in building trust. Open and honest communication allows players to clarify their intentions, share information, and reduce uncertainty. Through effective communication, players can develop shared expectations and establish a common understanding of the game being played.

External enforcement mechanisms can also foster trust in strategic interactions. When there are third-party institutions or mechanisms that can enforce agreements or penalize defectors, players are more likely to trust each other and engage in cooperative behavior. These mechanisms can include legal systems, contracts, reputation systems, or even social norms and conventions.

Real-World Applications

Reputation and trust have significant implications in a wide range of real-world scenarios. Let's consider a few examples:

Online Marketplaces: In platforms like Amazon or eBay, sellers with higher ratings and positive reviews have better reputations, leading to increased sales and trust from potential buyers. Trust is crucial in facilitating transactions between unknown parties in an online environment.

Business Alliances: When two companies enter into a strategic alliance, their decisions are influenced by the reputation and trustworthiness of each other. A positive reputation can lead to a deeper level of cooperation, more favorable terms, and shared benefits.

International Diplomacy: Countries build their reputation and trustworthiness through consistent compliance with international agreements and norms. Trust between nations can shape diplomatic negotiations and enhance cooperation on various issues, such as disarmament or environmental protection.

Exercise

Consider a scenario where two companies, A and B, are deciding whether to form a strategic alliance. Both companies are aware of their respective reputations in the industry. Company A has a strong reputation for delivering high-quality products and maintaining long-term relationships, while Company B has a reputation for being unreliable and engaging in unethical practices.

Discuss the potential impact of reputation on the decision-making process of both companies. How might it influence their willingness to trust each other and form a cooperative alliance? What strategies or actions could Company B undertake to overcome its negative reputation and build trust with Company A?

Conclusion

Reputation and trust are fundamental concepts in strategic interactions. A good reputation can act as a valuable asset, enhancing a player's credibility and opportunities for cooperation. Trust allows for the formation of alliances and agreements, leading to mutually beneficial outcomes. Building and maintaining reputation and trust requires consistent actions, repeated interactions, effective communication, and external enforcement mechanisms. Understanding and leveraging these concepts can lead to better outcomes in strategic interactions in various domains, from business alliances to international diplomacy.

Game Theory and Strategic Interaction in Politics and International Relations

In the realm of politics and international relations, strategic interaction plays a crucial role in decision-making processes. Game theory provides a powerful framework to analyze and understand the dynamics of these interactions. This section explores the application of game theory in political settings, from bargaining and competition to cooperation and conflict resolution.

Strategies and Tactics in Bargaining

Bargaining is a fundamental aspect of political negotiations, whether between countries, political parties, or interest groups. Game theory offers insights into strategies and tactics employed in these situations.

One classic game that captures the essence of bargaining is the *Ultimatum Game*. Imagine two players, A and B, who need to divide a sum of money. Player A proposes a split, and Player B can either accept or reject the offer. If the offer is accepted, the money is divided as proposed. However, if the offer is rejected, both players receive nothing.

Rationally, Player B should accept any positive offer, as rejecting it would result in a worse outcome for both. This assumption is based on the assumption of *rationality*, central to game theory. However, research has shown that B often rejects low offers due to a sense of fairness or a desire to punish unfair behavior.

In political bargaining, understanding this dynamics is essential. Decision-makers must carefully craft their proposals and consider potential reactions from their counterparts. By strategically aligning their interests and concessions, political actors can aim to achieve more favorable outcomes.

The Ultimatum Game: Fairness and Rationality

Let's delve deeper into the Ultimatum Game and its implications in politics. Consider a situation where two political parties are negotiating the distribution of resources. Party A proposes a division, and Party B must decide whether to accept or reject the offer.

To analyze this scenario, we can use payoff matrices. Let's assume there are two possible offers: a fair offer, where Party A proposes an equal split of resources, and an unfair offer, where Party A proposes a disproportionate division in their favor.

	Accept	Reject
Fair Offer	$\frac{1}{2}, \frac{1}{2}$	0, 0
Unfair Offer	0, 1	0, 0

In this matrix, the numbers represent the utilities for each outcome. For instance, if Party A proposes a fair offer and Party B accepts, both parties receive a utility of $\frac{1}{2}$. If Party B rejects the fair offer, both parties receive zero utilities.

The results demonstrate the importance of fairness considerations. Party A might be inclined to offer more evenly split resources to maximize the chances of acceptance by Party B. Conversely, Party B needs to weigh the potential gain by accepting an unfair offer against the perceived injustice.

Auctions: Bidding for Success

Another significant aspect of strategic interaction in politics is bidding processes, such as auctions for public contracts or political positions. Game theory provides a useful framework to study these processes and understand the tactics employed by participants.

Consider a sealed-bid auction, where multiple political actors submit bids for a desirable public contract. The highest bidder wins the contract but pays the price they bid.

To analyze this scenario, we can use the concept of *dominant strategies*. In game theory, a dominant strategy is one that yields the best outcome regardless of the actions taken by other players.

In a sealed-bid auction, each bidder must decide on their bid. The dominant strategy is to bid the maximum amount that the bidder is willing to pay for the contract. By doing so, they increase the chances of winning while avoiding overpaying if they win.

However, setting the right bid is challenging. Bidders must consider the likelihood of winning with a particular bid and the potential profit or benefit from securing the contract. Strategic thinking and accurate valuation become essential in determining the optimal bid.

Game Theory in Marketing and Consumer Behavior

Beyond political negotiations, game theory also offers valuable insights into strategic interaction in marketing and consumer behavior. Understanding how firms strategically position their products and how consumers make decisions is vital in this context.

One concept that relates to game theory in marketing is *branding*. Brands leverage their identity, reputation, and perceived value to gain a competitive advantage in the market.

Consider a scenario where two firms are competing in the same industry. Each firm can set either a high or low price for their product. By analyzing their competitors' potential strategies, firms can determine their optimal pricing strategy.

If either firm sets a high price while the other sets a low price, the one with the low price will capture a larger market share. However, if both firms set high prices, they might benefit from higher profit margins. This dilemma creates a strategic interaction where firms must carefully consider the actions of their competitors.

Additionally, firms use advertising as a strategic tool to influence consumer behavior. Advertising campaigns can create perceptions of quality, differentiate products from competitors, or evoke emotional responses.

Understanding the strategic nature of marketing and consumer behavior provides marketers with a framework to craft effective strategies, target specific segments, and gain a competitive advantage.

Reputation and Trust in Strategic Interactions

Trust and reputation play crucial roles in strategic interactions. Establishing a good reputation and fostering trust can lead to more favorable outcomes, while a tarnished reputation can have lasting negative effects.

Consider a political leader seeking to negotiate a diplomatic agreement. The leader's reputation for keeping their promises and acting in good faith can influence how other countries perceive their offers and intentions.

In game theory, this aspect is captured by *repeated games*. In a repeated interaction scenario, players have the opportunity to build a reputation based on past actions. This reputation, in turn, influences how others interact with the player in subsequent iterations of the game.

Maintaining a trustworthy and reliable reputation can enhance a political leader's bargaining power and increase the chances of reaching mutually beneficial agreements.

Game Theory and Strategic Interaction in Politics and International Relations: Key Takeaways

Game theory offers a powerful toolkit to analyze strategic interactions in politics and international relations. By understanding the strategies, tactics, and dynamics involved, decision-makers can make more informed choices, negotiate effectively, and achieve favorable outcomes. Important concepts in this domain include bargaining strategies, fairness considerations, auctions as bidding processes, reputation and trust, and their implications in marketing and consumer behavior. Applying game theory to political and international contexts enables a deeper understanding of the complexities involved and provides a framework for decision-making.

Exercises

1. Reflecting on the Ultimatum Game, consider a real-world scenario where two political parties are negotiating the allocation of public resources. Discuss the factors that could influence the acceptance or rejection of an offer from the perspective of both parties.

2. Choose a current political negotiation or conflict resolution process and analyze it using game theory. Identify the players, their strategies, and the potential outcomes. Discuss the influence of factors such as reputation, trust, and fairness on the decision-making process.

3. Explore a case study where branding and reputation management played a significant role in shaping consumer behavior and market outcomes. Analyze the strategies employed by the brand to gain a competitive advantage and the impact on consumer perception.

4. Investigate a real-world scenario where trust and reputation were pivotal in shaping diplomatic relationships between countries. Discuss the role of past interactions, promises kept or broken, and their influence on future negotiations.

5. Research current debates or controversies related to the use of game theory in politics and international relations. Present arguments for and against the application of game theory in these contexts and discuss potential ethical implications.

Additional Resources

Books

- Dixit, A., and Nalebuff, B. (2008). *The Art of Strategy: A Game Theorist's Guide to Success in Business and Life.*

- Schelling, T. C. (1980). *The Strategy of Conflict.*

- Serrato, F. (2017). *Game Theory in Communication Networks: Cooperative Resolution of Interactive Networking Scenarios.*

Articles

- Axelrod, R., and Hamilton, W. D. (1981). *The Evolution of Cooperation.* Science, 211(4489), 1390-1396.

- Fehr, E., and Gachter, S. (2002). *Altruistic Punishment in Humans.* Nature, 415(6868), 137-140.

- Kreps, D. M. (1990). *Corporate Culture and Economic Theory.* In Perspectives on Positive Political Economy (pp. 90-143). Cambridge University Press.

Websites

- Stanford Encyclopedia of Philosophy: `https://plato.stanford.edu/entries/game-theory/`

- Game Theory Society: `https://www.gametheorysociety.org/`

Game Theory in Financial Markets

Game Theory and Financial Decision-Making

Market Efficiency and Game Theory

In this section, we will explore the relationship between market efficiency and game theory. Market efficiency refers to the degree to which market prices accurately reflect all available information. On the other hand, game theory provides a framework for understanding strategic interactions and decision-making in various market settings. By combining these two concepts, we can better understand the dynamics of financial markets and how game theory can be applied to enhance market efficiency.

Efficient Market Hypothesis

Before we dive into the connection between market efficiency and game theory, let's briefly discuss the Efficient Market Hypothesis (EMH). EMH is a fundamental principle in finance that states that financial markets are efficient and that it is impossible to consistently achieve higher than average returns through trading strategies.

The EMH is based on the assumption that all market participants have access to the same information and act rationally. According to this hypothesis, prices in financial markets reflect all available information, making it difficult for any individual investor to gain an advantage consistently.

Game Theory and Market Efficiency

Game theory provides a valuable tool for understanding the strategic interactions that occur in financial markets. It allows us to model the decision-making process of various market participants, such as investors, traders, and market makers.

In the context of market efficiency, game theory helps us analyze the behavior of investors and how they respond to new information. In an efficient market, investors aim to maximize their expected utility by incorporating all available information into their trading decisions. Game theory helps us understand how investors form expectations, react to price changes, and adjust their trading strategies based on the actions of others.

One of the fundamental concepts in game theory is the Nash equilibrium, which is a stable state of a game in which each player chooses their best strategy, given the strategies chosen by others. In an efficient market, the Nash equilibrium captures the idea that all market participants are acting rationally and optimizing their decision-making based on available information.

Market Anomalies and Behavioral Biases

While market efficiency is an important concept, it is essential to recognize that market anomalies and behavioral biases can lead to deviations from efficiency. Market anomalies refer to situations where prices deviate from their fundamental values, creating opportunities for investors to profit from mispricing.

Behavioral biases, on the other hand, result from psychological factors that influence decision-making. Investors may exhibit irrational behaviors, such as overreaction to news or herd mentality, which can create market inefficiencies.

Game theory helps us understand these deviations from efficiency by modeling the strategic interactions between rational and irrational investors. By considering the actions and beliefs of both types of investors, we can gain insights into the underlying dynamics that lead to market anomalies.

Real-World Examples

To illustrate the connection between market efficiency and game theory, let's consider a real-world example. Suppose there is a company, ABC Corp, that is about to release its quarterly earnings report. Investors have access to different pieces of information, some of which suggest positive earnings surprises, while others indicate negative ones.

In this scenario, game theory can help us analyze how investors respond to this information and how it impacts market efficiency. Rational investors may use game theory principles to assess the actions and beliefs of others to determine the most optimal trading strategy.

For instance, if most investors are optimistic about the earnings report, we might expect the market price of ABC Corp's stock to increase. However, if some investors have access to negative information, they may choose to sell their shares, anticipating a decline in the stock price.

Understanding the strategic interactions between these rational investors and those who exhibit behavioral biases can help us identify market inefficiencies and potential profit opportunities.

Resources and Further Reading

If you're interested in exploring market efficiency and game theory further, here are some recommended resources:

1. *A Random Walk Down Wall Street* by Burton G. Malkiel
2. *Thinking, Fast and Slow* by Daniel Kahneman
3. *Game Theory for Applied Economists* by Robert Gibbons
4. *An Introduction to Game Theory* by Martin J. Osborne and Ariel Rubinstein

These resources provide valuable insights into market efficiency, behavioral biases, and their connection to game theory.

Key Takeaways

- Market efficiency refers to the degree to which market prices accurately reflect all available information.

- Game theory helps us understand strategic interactions and decision-making in financial markets.

- The Efficient Market Hypothesis states that financial markets are efficient, and it is difficult to consistently achieve above-average returns.

- Game theory helps us analyze investor behavior in an efficient market and understand deviations from efficiency caused by market anomalies and behavioral biases.

- Real-world examples demonstrate how game theory can be applied to enhance market efficiency.

Exercise

Consider a scenario where a new piece of information is released, suggesting that a company's CEO is involved in a financial scandal. Using game theory principles, analyze how investors might react to this information and its potential impact on market efficiency.

Hint: Consider the strategic interactions between rational investors and those influenced by behavioral biases.

Behavioral Finance and Game Theory

In this section, we will explore the fascinating intersection between behavioral finance and game theory. While traditional finance assumes that individuals are rational decision-makers who always act in their own best interest, behavioral finance recognizes that people are not always rational and are influenced by psychological biases. Game theory, on the other hand, provides a framework for analyzing strategic interactions between rational decision-makers. By combining these two disciplines, we can gain a deeper understanding of how psychological factors affect financial decision-making and explore the implications for investment strategies.

The Role of Psychology in Finance

Behavioral finance is a field of study that examines how psychological biases and heuristics influence financial decision-making. It challenges the traditional notion of the "rational investor" and highlights the importance of understanding human behavior in financial markets.

One of the key insights from behavioral finance is that individuals often deviate from rationality due to cognitive biases. These biases can lead to mispricing of assets and irrational investment decisions. For example, the availability bias refers to the tendency to make judgments based on easily recalled information, leading investors to overweight recent or salient events. This can result in herd behavior and the formation of speculative bubbles in financial markets.

Another important bias is the anchoring bias, where individuals rely too heavily on an initial piece of information when making subsequent decisions. This can cause investors to hold onto losing positions for too long or to set unrealistic expectations for future returns.

Understanding these biases is crucial for designing effective investment strategies and risk management techniques. Game theory provides a valuable framework for studying how these biases can affect the behavior of investors in strategic interactions.

Game Theory Applications in Financial Decision-Making

Game theory allows us to analyze interactions between multiple decision-makers, accounting for their rationality, motivations, and strategies. In the context of financial decision-making, game theory can help us understand how various market participants interact and how their strategies and decisions affect overall market outcomes.

One of the fundamental concepts in game theory is the Nash equilibrium, which represents a stable state where no player can unilaterally improve their outcome. In financial markets, identifying Nash equilibria can help us understand the pricing dynamics and behavior of market participants.

For example, consider a stock market where investors can either buy or sell a particular stock. Each investor's decision depends on their perceptions of the stock's value and their expectations of other investors' actions. By modeling this interaction as a game using game theory, we can analyze the equilibrium outcomes and predict the behavior of investors in different market conditions.

Another application of game theory in finance is the analysis of strategic interactions between financial institutions. For instance, in a competitive lending market, banks must decide how much risk to take on while maximizing their profits. Using game theory, we can study how banks' lending strategies interact and assess the stability of the resulting market outcomes.

Real-World Examples

To better understand the practical implications of behavioral finance and game theory in finance, let's explore some real-world examples.

1. **The Prisoner's Dilemma in Financial Regulation**: The Prisoner's Dilemma is a classic game in which two players have to decide whether to cooperate or defect. In the context of financial regulation, this dilemma arises when multiple banks must decide whether to take on excessive risk or adopt more conservative practices. If one bank chooses to take on more risk, it can potentially earn higher returns, but at the risk of destabilizing the entire financial system. However, if all banks simultaneously pursue conservative strategies, they may miss out on potential profits. Understanding the dynamics of this dilemma can help regulators design effective policies to prevent systemic risks.

2. **Herding Behavior in Stock Markets**: Behavioral biases can lead to herding behavior in stock markets, where investors follow the decisions of others instead of conducting independent analysis. This phenomenon can result in market inefficiencies and mispricing of assets. By using game theory to model this interaction, we can explore the conditions under which herding behavior emerges and its impact on market stability.

3. **Bargaining Strategies in Mergers and Acquisitions**: Mergers and acquisitions involve complex negotiations and strategic interactions between companies. By incorporating game theory into the analysis, we can examine the bargaining power of each party and predict the likely outcomes of these transactions. Understanding the strategic behavior of companies in mergers and acquisitions can help investors make informed decisions regarding their investment portfolios.

Limitations and Future Directions

While behavioral finance and game theory provide valuable insights into financial decision-making, they also have some limitations.

First, behavioral finance relies on the assumption that individuals exhibit consistent patterns of behavior over time. However, individuals can learn from their experiences and adapt their decision-making, which may limit the predictability of certain biases.

Second, game theory assumes that all decision-makers are rational and have complete information, which may not always hold in practice. In reality, individuals may have limited information or act based on incomplete analysis.

Future research in behavioral finance and game theory could explore how these theories can be integrated with other disciplines, such as artificial intelligence and machine learning, to improve our understanding of financial markets. Additionally, the ethical implications of using these theories in practice should be carefully considered to ensure fair and transparent financial outcomes.

Further Resources

To delve deeper into the fascinating world of behavioral finance and game theory, consider exploring the following resources:

- *Thinking, Fast and Slow* by Daniel Kahneman: This book by Nobel laureate Daniel Kahneman provides a comprehensive overview of the cognitive biases and heuristics that influence human decision-making.

- *Game Theory: Analysis of Conflict* by Roger B. Myerson: This book offers a detailed introduction to game theory and its applications in various fields, including finance.

- *Misbehaving: The Making of Behavioral Economics* by Richard H. Thaler: In this book, Richard H. Thaler, a pioneer in the field of behavioral economics, shares his insights and experiences in challenging the traditional assumptions of economic rationality.

- *The Strategy of Conflict* by Thomas C. Schelling: This classic work explores the role of strategic thinking and game theory in resolving conflicts and negotiating outcomes.

By studying behavioral finance and game theory, we can gain a deeper understanding of the complexities of financial decision-making and develop strategies to navigate the ever-changing landscape of the financial markets. So, buckle up and get ready to play smart, not hard!

Game Theory and Asset Pricing Models

In this section, we will explore the fascinating intersection between game theory and asset pricing models. Asset pricing models play a crucial role in finance, helping investors evaluate the value of financial assets and make informed investment decisions. Game theory, on the other hand, provides a strategic framework for analyzing decision-making in competitive situations. By combining these two fields, we can gain a deeper understanding of how players in financial markets interact and determine asset prices.

Background

Before diving into the connection between game theory and asset pricing models, let's first lay down some important concepts from both areas.

Asset Pricing Models Asset pricing models aim to determine the fair value of financial assets, such as stocks, bonds, and derivatives. These models consider various factors, including risk, return, interest rates, and market conditions, to estimate the intrinsic value of an asset. They provide a framework for comparing the value of an asset with its current market price, allowing investors to identify potential undervalued or overvalued assets.

Some well-known asset pricing models include the Capital Asset Pricing Model (CAPM), the Arbitrage Pricing Theory (APT), and the Black-Scholes-Merton model for option pricing.

Game Theory Game theory, developed by mathematician John von Neumann and economist Oskar Morgenstern, studies the strategic interaction between individuals or groups in competitive situations. It provides a mathematical framework for analyzing decision-making, considering the choices made by different players and the potential outcomes of their actions.

Key concepts in game theory include players, strategies, payoffs, and equilibria. Players are the decision-makers in a game, while strategies represent the choices available to them. Payoffs represent the outcomes or rewards associated with different strategy combinations. Equilibria, such as Nash equilibrium, occur when no player can unilaterally improve their payoff by changing their strategy.

Bringing Game Theory to Asset Pricing Models

Now that we have a basic understanding of asset pricing models and game theory, let's explore how these two fields intertwine.

Asset pricing models traditionally assume that investors are rational and act in their own best interest. However, game theory acknowledges that decision-makers may also consider the actions and strategies of others when making choices.

By applying game theory to asset pricing models, we can capture the strategic interactions that occur in financial markets. This allows us to better understand how investors' actions and expectations influence asset prices.

One way game theory is incorporated into asset pricing models is through the concept of market microstructure. Market microstructure refers to the mechanisms, rules, and institutions that govern the trading of financial assets.

Game-theoretic models of market microstructure take into account the strategic behavior of market participants, such as investors and market makers. These models analyze how investors' trading decisions and market makers' pricing strategies interact to determine asset prices. They consider factors like order flow, bid-ask spreads, and liquidity provision.

Additionally, game theory can shed light on the formation of bubbles and speculative episodes in asset markets. By studying how players' expectations and actions influence each other, we can gain insights into the dynamics behind price deviations from fundamental values. These deviations can be attributed to herd behavior, information cascades, and irrational exuberance.

Real-Life Applications

To see game theory and asset pricing models in action, let's consider a real-life example - the initial public offering (IPO) of a tech company.

When a tech company decides to go public and offer its shares to the public, it needs to determine the IPO price. This price will impact the company's valuation, as well as the returns for both institutional and retail investors.

Game theory can help us understand the strategic considerations involved in this process. The tech company wants to maximize its valuation and raise as much capital as possible, while investors want to buy shares at a price that offers a favorable risk-return tradeoff.

The company faces a coordination problem - it needs to set a price that attracts enough investors to ensure a successful IPO. This requires anticipating the investors' strategies and valuations. On the other hand, investors need to evaluate the company's prospects and the likelihood of receiving an allocation of shares at the IPO price.

Game-theoretic models can analyze these dynamics and help both the company and investors make better decisions. By considering various factors, such as demand uncertainty, information asymmetry, and the behavior of other market participants, we can develop models that capture the strategic interactions and provide insights into the IPO pricing process.

Challenges and Caveats

While incorporating game theory into asset pricing models offers valuable insights, there are also challenges and caveats to be aware of.

One challenge is the complexity of modeling strategic interactions in financial markets. Financial markets involve a large number of participants with different goals and strategies, making it difficult to capture the full dynamics of their interactions. Simplifying assumptions are often necessary to make the models tractable.

Another challenge is the assumption of rationality. Game theory assumes that players are rational, meaning they always make the best decisions given their available information. However, numerous studies have shown that human behavior deviates from rationality in practice. Behavioral biases, emotions, and cognitive limitations can influence decision-making in unpredictable ways.

Additionally, game-theoretic models in finance are often based on simplified assumptions about market structure and participants' preferences. Real-world markets can be much more complex, with factors like asymmetric information, transaction costs, and regulatory constraints affecting outcomes.

Conclusion

In summary, game theory offers a powerful framework for understanding the strategic interactions in financial markets, complementing traditional asset pricing models. By incorporating game-theoretic concepts into asset pricing models, we can gain deeper insights into the dynamics of asset prices, market microstructure, and investor behavior.

However, it is important to recognize the challenges and limitations of these models. Game theory provides a useful starting point but should be supplemented with other tools and approaches to capture the full complexity of financial markets.

The connection between game theory and asset pricing models opens up exciting avenues for research and exploration. As financial markets continue to evolve, this interdisciplinary approach will be essential for understanding and navigating the intricacies of modern finance.

Exercises

1. Consider a scenario where two investors are valuing a stock before its IPO. Investor A believes the stock will have a high valuation and is willing to pay a higher price, while Investor B is more cautious and believes the stock is overvalued. Use game theory to analyze the potential outcomes and strategies of both investors.

2. Research and analyze a real-world example of a financial market where game theory principles can provide insights into asset pricing. Explain how game theory concepts can be applied to understand the strategic interactions in that market.

3. Discuss the limitations of incorporating game theory into asset pricing models. How can behavioral biases and market complexities influence the accuracy of these models?

4. Critically evaluate the assumption of rationality in game theory and asset pricing models. How does the departure from rationality impact the outcomes and predictions of these models in practice?

5. Explore the role of game theory in high-frequency trading and algorithmic trading strategies. How do these strategies incorporate strategic interactions and affect asset prices in financial markets?

Further Reading

1. Tirole, Jean. "The Theory of Corporate Finance." Princeton University Press, 2006.
2. Fudenberg, Drew, and Jean Tirole. "Game Theory." MIT Press, 1991.
3. Shleifer, Andrei. "Inefficient Markets: An Introduction to Behavioral Finance." Oxford University Press, 2000.
4. Jackson, Matthew O. "Social and Economic Networks." Princeton University Press, 2008.
5. Dixit, Avinash K., and Barry J. Nalebuff. "Thinking Strategically: The Competitive Edge in Business, Politics, and Everyday Life." W. W. Norton & Company, 1993.

Game Theory in Portfolio Selection

In this section, we will explore how game theory can be applied to the field of portfolio selection in financial markets. Portfolio selection is the process of choosing the optimal mix of assets to include in an investment portfolio based on an individual's risk tolerance and expected return.

Background

Portfolio selection is a complex task because multiple factors need to be taken into consideration, including the risk and return trade-off, correlation between assets, and the investor's risk aversion. Traditionally, portfolio selection models have focused on mean-variance analysis, where the goal is to maximize the expected return while minimizing the portfolio's variance or risk.

However, game theory offers an alternative approach to portfolio selection that takes into account the strategic interactions between investors. By modeling investment decisions as a game, where each player's objective is to maximize their own utility, we can gain insights into how different investment strategies and behaviors might impact portfolio selection.

Principles of Game Theory in Portfolio Selection

Game theory provides a framework for analyzing the strategic behavior of investors in portfolio selection. Here are some key principles of game theory that can be applied to this context:

1. Players: In portfolio selection, the players are the individual investors or fund managers who are making investment decisions. Each player aims to maximize their own utility, which is typically measured by a combination of expected return and risk.

2. Strategies: Players have different investment strategies and tactics they can employ, such as diversification, active management, or passive investing. These strategies can be thought of as the player's actions in the game.

3. Payoffs: The payoffs in portfolio selection are the returns and risks associated with different investment choices. Each player's payoff depends on the returns they earn and the risk they bear, as well as the strategic behavior of other players.

4. Information: Players have access to different types of information, including market data, news, and their own risk preferences. The amount and quality of information available to each player can influence their investment decisions.

5. Equilibrium: In game theory, equilibrium represents a stable state where no player has an incentive to change their strategy unilaterally. In the context of portfolio selection, equilibrium can be achieved when each player's strategy is optimal given the strategies of other players.

Models of Game Theory in Portfolio Selection

There are several game-theoretic models that have been proposed for portfolio selection. Here, we will discuss two popular models: the mean-variance model with strategic interaction and the evolutionary game model.

1. Mean-Variance Model with Strategic Interaction: This model extends the traditional mean-variance framework by incorporating game-theoretic elements. Players aim to find the optimal asset allocation that maximizes their expected return while minimizing their risk, taking into account the actions of other players. Nash equilibrium, a concept from game theory, can be used to identify the optimal equilibrium asset allocation strategy.

2. Evolutionary Game Model: This model applies concepts from evolutionary biology to portfolio selection. Players are represented by different investment strategies, and their performance over time determines their "fitness". Strategies with higher fitness have a higher

probability of being adopted by other players in the future. Through a process of natural selection, the population of investment strategies evolves, potentially leading to the emergence of a dominant strategy or a stable equilibrium.

Real-World Applications

Game theory has found practical applications in portfolio selection. Here are a few examples:
 1. Market Microstructure: Game theory can be used to model the strategic interactions between investors, brokers, and market makers in financial markets. By understanding how different players behave and react to changes in market conditions, investors can make more informed decisions when constructing their portfolios.
 2. Risk Management: Game theory can help investors assess potential risks associated with their investment decisions. By modeling the interactions between different market participants, such as the effects of herding behavior or speculation, investors can better understand and manage their exposure to different types of risks.
 3. Impact of News and Sentiment: Game theory can be used to analyze how news and market sentiment influence investors' decision-making. By incorporating the strategic reactions of investors to new information, portfolio managers can adjust their strategies accordingly, taking advantage of market inefficiencies.

Caveats and Considerations

When applying game theory to portfolio selection, there are several caveats and considerations to keep in mind:
 1. Simplifying Assumptions: Game-theoretic models often rely on simplifying assumptions about the behavior and rationality of market participants. These assumptions may not always accurately reflect the complex nature of financial markets.
 2. Information Assymetry: In practice, investors may have access to different levels and quality of information. Assumptions about symmetric information among players may not hold, which can affect the effectiveness of game-theoretic models.
 3. Computational Complexity: Game-theoretic models can be computationally intensive, especially when dealing with large portfolios

or complex investment strategies. Implementing these models may require advanced computational tools and expertise.

Conclusion

Game theory offers a fresh perspective on portfolio selection in financial markets, allowing investors to consider the strategic interactions between market participants. By incorporating game-theoretic models into the decision-making process, investors can make more informed and strategic choices when constructing their portfolios. However, it is important to acknowledge the limitations and assumptions of these models and adapt them to the specific context and requirements of the investment environment.

Auctions and Game Theory in Finance

In the fast-paced world of finance, auctions play a crucial role in determining prices, allocating resources, and facilitating transactions. Auction theory, a branch of game theory, provides us with a powerful toolkit to analyze and understand how auctions work and how participants can strategically navigate these markets. In this section, we will dive into the fascinating realm of auctions and explore the intricate connections between game theory and finance.

Introduction to Auctions

Let's start by understanding what exactly an auction is. An auction is a market mechanism in which buyers compete to acquire a good or service through a bidding process. The seller establishes specific rules and conditions, and participants place bids according to these rules. The highest bidder, also known as the winner, secures the item and pays the price they bid.

Auctions have been around for centuries, and they have evolved to suit different contexts and purposes. In finance, auctions are commonly used to sell a wide range of financial assets, such as government bonds, treasury bills, stocks, and derivatives. Auctions also play a critical role in IPOs (Initial Public Offerings), where shares of a company are sold to the public for the first time.

Types of Auctions

Different types of auctions exist, each with its unique characteristics and bidding rules. Let's explore a few of the most common types:

1. English Auction (Open Ascending Auction): In this type of auction, the auctioneer starts with a low initial price and gradually increases it. Bidders indicate their willingness to pay by calling out higher amounts until no one is willing to bid any further. The highest bidder wins the item and pays the price they bid.

2. Dutch Auction (Open Descending Auction): In a Dutch auction, the auctioneer starts with a high initial price and progressively lowers it until a bidder is willing to accept the price. The first bidder to accept wins the item.

3. First-Price Sealed-Bid Auction: In this type of auction, bidders submit sealed bids without knowing what others have bid. The highest bidder wins the item and pays the price they bid.

4. Second-Price Sealed-Bid Auction (Vickrey Auction): In a second-price sealed-bid auction, bidders submit sealed bids, and the highest bidder wins the item. However, the price paid by the winner is not their bid but rather the amount of the second-highest bid. This encourages bidders to bid truthfully, as they have no incentive to shade their bids.

Auction Strategies and Tactics

To succeed in an auction, participants often employ various strategies and tactics to maximize their chances of winning while minimizing the price paid. Let's explore a few common strategies:

1. Bid shading: Bidders might deliberately shade their bids to avoid revealing their true valuation of the item. By bidding slightly below their maximum willingness to pay, they increase the likelihood of winning at a lower price.

2. Sniping: Bidders can employ a strategy called sniping, where they place a bid at the very last moment in an attempt to outbid other participants without giving them a chance to respond effectively. This strategy aims to exploit the element of surprise and prevent counterbidding.

3. Bid incrementing: Some bidders may increment their bids by small amounts rather than revealing their maximum willingness to pay right away. This tactic helps them test the resolve of their competitors and potentially discourage them from bidding higher.

4. **Collusion:** In certain cases, bidders might collude or form a cartel to manipulate the auction outcome in their favor. This strategy is illegal in many jurisdictions, as it undermines fair competition and distorts market outcomes.

Auctions in Financial Markets

Auctions play a vital role in financial markets, where they facilitate price discovery and ensure efficient allocation of scarce resources. Here are a few real-world examples:

1. **Treasury Bond Auctions:** Governments use auctions to issue and sell bonds to investors. These auctions help determine the market interest rates on government debt and provide a benchmark for other interest rates in the economy.

2. **Stock Exchange Auctions:** Stock exchanges conduct auctions for IPOs and other corporate actions such as rights offerings and mergers. These auctions enable companies to raise capital and investors to acquire shares of newly listed firms.

3. **Derivatives Auctions:** Financial derivatives, such as options and futures contracts, are often traded through auctions. These auctions allow investors to buy or sell derivative contracts at fair market prices and hedge their exposures.

4. **Foreclosure Auctions:** In the unfortunate event of a borrower defaulting on their mortgage, foreclosure auctions are conducted to sell the property and recover the outstanding loan amount. Interested buyers bid on the property during such auctions.

5. **Online Auction Platforms:** With the rise of e-commerce, online auction platforms like eBay and Amazon Auctions have gained popularity. These platforms enable individuals and businesses to buy and sell a wide range of products through competitive bidding.

Auction Design and Efficiency

Auction design is crucial to ensure efficiency and fairness in market outcomes. In finance, auction designers focus on two key objectives: revenue maximization for the seller and social welfare maximization for society. The design of an auction depends on the characteristics of the item being sold, market conditions, and the desired outcome.

Common Auction Problems and Solutions

Auctions are not immune to challenges and issues. Let's explore a few common problems and possible solutions:

1. **Winner's Curse:** In certain auctions, the winner may end up overpaying if their valuation of the item was overly optimistic. The winner's curse can be mitigated by conducting auctions with more information sharing or by employing a second-price sealed-bid auction format.

2. **Bidder Collusion:** Collusion between bidders can compromise market competition and lead to unfair outcomes. Auction organizers should implement mechanisms to detect and deter collusion, such as strict anti-collusion regulations and independent monitoring.

3. **Information Asymmetry:** When bidders have different levels of information about the item being auctioned, market efficiency can suffer. Auction designers can address this issue by ensuring transparency and providing relevant information to all participants.

Real-World Example: Spectrum Auctions

A fascinating real-world application of auction theory in finance is the auctioning of wireless spectrum licenses. Governments around the world use auctions to allocate spectrum rights to telecommunications companies. These auctions are complex, high-stakes events that require careful design to ensure efficient allocation and fair competition.

Telecom companies bid for the rights to use specific frequency bands for wireless communication services. The auction design often involves multiple rounds, sealed bids, and simultaneous ascending-clock auctions to allocate different blocks of spectrum. The goal is to maximize revenue for the government while enabling competition and efficient use of the spectrum.

Spectrum auctions provide an excellent case study for the application of game theory in finance. Bidders must navigate a strategic landscape, considering not only the value they derive from the spectrum but also the potential actions and responses of their competitors.

Conclusion

Auctions are dynamic and exciting markets that have a profound impact on finance and other sectors of the economy. By employing game theory, we can better understand the strategic interactions between bidders and design auctions that maximize efficiency and welfare.

In this section, we explored the fundamentals of auctions, different auction types, strategies employed by participants, and their application in financial markets. We also examined the importance of auction design, common problems, and real-world examples. Auctions are a fascinating area of study where finance meets game theory, offering endless opportunities for research, innovation, and understanding market dynamics.

Option Pricing and Game Theory

In the exciting world of finance, where fortunes are made and lost in the blink of an eye, understanding option pricing is crucial. Options are financial derivatives that give the holder the right, but not the obligation, to buy or sell an underlying asset at a predetermined price within a specific timeframe. Game theory provides a powerful framework for analyzing and pricing these options, taking into account the strategic behavior of market participants.

Background: Option Pricing

Before delving into the game-theoretic aspects of option pricing, let's first review some foundational concepts. The Black-Scholes model, developed by economists Fischer Black and Myron Scholes in 1973, revolutionized the field of quantitative finance. This model provides a mathematical framework for determining the fair price of options.

The Black-Scholes model assumes that the price movement of the underlying asset follows a geometric Brownian motion and that market participants are risk-neutral. By using a combination of stochastic calculus and partial differential equations, the model provides an equation, known as the Black-Scholes equation, for calculating the price of a European option.

However, the Black-Scholes model has some limitations. It assumes constant volatility, which does not always hold in real-world markets. It also assumes continuous trading and does not account for transaction costs or market frictions. These shortcomings have motivated researchers to explore alternative approaches, including game theory.

Game Theory and Option Pricing

Game theory is a mathematical framework for studying decision-making in situations where multiple agents, or players, interact strategically. In the context of option pricing, game theory allows us to model how market participants strategically set their buying and selling strategies to maximize their payoffs.

One approach rooted in game theory is the concept of incomplete markets. Incomplete markets recognize that not all assets can be perfectly replicated using combinations of other assets. This can lead to market inefficiencies and potential bargaining power between buyers and sellers of options.

In a game-theoretic framework, option pricing can be viewed as a strategic interaction between two types of players: option buyers and option sellers. Option buyers aim to maximize their payoff by anticipating future price movements, while option sellers seek to optimize their own risk exposure.

One popular game-theoretic model for option pricing is the binomial options pricing model. This model assumes that the underlying asset can only take two possible values at each time step, creating a binomial tree. By recursively calculating option prices at each step, we can obtain the fair value of the option.

Strategic Considerations in Option Pricing

In the game of options, both buyers and sellers need to carefully consider their strategies to achieve their desired outcomes. Here are a few strategic considerations relevant to option pricing:

1. Information Advantage: Traders with superior information or analysis may have an edge in option pricing. For example, if an option buyer has access to non-public information, they may be able to profit by correctly predicting future price movements.

2. Market Liquidity: The availability of buyers and sellers in the market can impact option prices. Illiquid markets may result in wider bid-ask spreads, making it more expensive for buyers to enter into options contracts.

3. Risk Management: Option sellers are exposed to potential losses if the underlying asset moves against their position. They need to assess and manage their risk exposure to protect their capital.

4. Hedging Strategies: Both buyers and sellers can use hedging strategies to reduce risk. For example, option sellers can hedge their exposure by taking offsetting positions in the underlying asset.

Real-World Examples

To illustrate the application of game theory in option pricing, let's consider a real-world example involving a technology company, TechCo, and an investor, John.

TechCo's stock is currently trading at 100 per share, and John believes that the stock price will increase in the next month. E per share.

Simultaneously, TechCo's management expects the stock price to decrease due to a negative earnings announcement. They are willing to sell call options to generate additional income while mitigating their downside risk.

In this scenario, both John and TechCo engage in strategic decision-making. John analyzes the risks and potential rewards of purchasing the call option, considering his information advantage and expectations of future price movements. TechCo assesses the impact of selling call options on their stock price and the overall risk exposure of their portfolio.

By applying game theory, we can model their strategic interactions and derive the fair price of the call option considering their preferences, expectations, and strategies.

Conclusion

Option pricing is a fascinating field where finance and game theory merge. By understanding the strategic interactions between option buyers and sellers, we can better evaluate the fair value of options and make informed investment decisions.

In this section, we explored how game theory provides a framework for analyzing option pricing. We discussed the foundational concepts of option pricing, the limitations of the Black-Scholes model, and the role of game theory in addressing these limitations. We also examined strategic considerations in option pricing, such as information advantage, market liquidity, risk management, and hedging strategies.

Understanding the game-theoretic aspects of option pricing empowers investors to navigate the complex world of financial markets and make sound investment decisions. Whether you're a seasoned trader or just starting, incorporating game theory into your investment toolkit can give you a competitive edge.

Now, armed with the knowledge of option pricing and game theory, let's explore how social networks and game theory intersect in the next section. We'll delve into social network analysis, centrality, and the diffusion of ideas through the lens of game theory. Get ready for some mind-blowing insights into how our interconnected world shapes our decision-making processes!

Game Theory in Risk Management and Insurance

Risk management and insurance are integral parts of our modern society. We face various risks every day, from health issues to accidents, natural disasters, and financial losses. Game theory offers a valuable framework to analyze and manage risks in these fields by considering the interactions and strategic decision-making of different parties involved. In this section, we will explore how game theory can be applied to risk management and insurance, providing insights into decision-making, optimal strategies, and overcoming challenges in these areas.

Risk Management: Identifying and Assessing Risks

Before diving into game-theoretic concepts, it's essential to understand risk management fundamentals. Risk management involves identifying, assessing, and prioritizing potential risks and developing strategies to minimize their adverse effects.

Game theory complements risk management by analyzing interactions between different stakeholders and their strategies in risk-related situations. It allows us to estimate the potential impact of decisions made by individuals or organizations while considering the potential actions of others.

Consider a scenario where a company wants to minimize the risk of a product launch failure. By applying game theory, the company can analyze the potential decisions of competitors, customers, and suppliers. This analysis enables the company to adjust its strategies, mitigate risks, and gain a competitive advantage.

Insurance: A Game of Risk Sharing

Insurance is a mechanism designed to transfer individual risks to a collective pool, reducing the financial impact of unexpected events. Game theory plays a vital role in understanding the dynamics between insurance providers and policyholders, as well as designing optimal insurance contracts that satisfy both parties.

In the realm of insurance, game theory helps address the issue of adverse selection, where policyholders with higher risks are more likely to seek insurance coverage. This can result in higher premiums for everyone, making insurance less affordable.

To mitigate adverse selection, insurance companies can use game-theoretic models to design contracts that incentivize policyholders to reveal their true risk profiles. By offering appropriate coverage options and premiums, insurance companies can encourage low-risk individuals to disclose accurate information while discouraging high-risk individuals from hiding their risks.

Game-Theoretic Models in Risk Management and Insurance

Game-theoretic models provide a structured approach to analyze risk management and insurance problems. These models typically involve multiple players with conflicting interests and uncertain outcomes. Let's explore some common game-theoretic models used in risk management and insurance:

1. **The Principal-Agent Model:** In insurance, the principal-agent problem arises when an insurer (principal) delegates decision-making to an insured individual (agent) who may not act in the principal's best interests. The insurer aims to align the agent's actions with their objectives, maintaining a balance between risk-sharing and minimizing opportunistic behavior.

   For example, an insurance company wants to incentivize the insured party to take necessary precautions to reduce risks. By designing insurance policies that link premiums to risk reduction efforts, the insurer can create a game where the insured's optimal strategy aligns with the insurer's goals.

2. **The Market for Lemons:** The market for lemons is a classic game-theoretic model used to analyze asymmetric information in insurance and risk management. It explores scenarios where sellers possess more information about the quality or risk of a product than buyers, leading to market inefficiencies.

   Applying this concept to insurance, policyholders may have more information about their risk profiles than insurance companies. This information asymmetry can lead to adverse selection, where high-risk individuals are more likely to seek insurance coverage. Insurance companies can mitigate asymmetric information problems by using risk assessment tools, offering different coverage options, or utilizing risk-sharing mechanisms.

3. **Collaborative Risk Management Games:** Collaborative risk management games involve multiple stakeholders who need to cooperate to achieve risk reduction. These games simulate situations where players must balance their individual benefits against collective risk reduction.

   For instance, in a community threatened by natural disasters, residents can engage in a game where they decide whether or not to invest in protective measures such as building dykes or early warning systems. By analyzing the potential strategies of residents and their interactions, game theory helps identify cooperative strategies that maximize risk reduction for the entire community.

Real-World Applications: Using Game Theory in Risk Management and Insurance

Game theory has found practical applications in risk management and insurance across various industries. Let's explore a few real-world examples:

1. **Health Insurance:** Health insurance companies face challenges in designing coverage plans that account for individuals' varying health risks. Game theory helps insurers estimate the risk profiles

of policyholders and develop contracts that balance individual coverage needs and affordability.

2. **Reinsurance:** Reinsurance companies provide coverage for primary insurers against catastrophic losses. Game theory helps in determining optimal reinsurance strategies, including risk sharing and pricing, considering the potential actions of primary insurers and the extent of potential losses.

3. **Cyber Risk Management:** With the growing prevalence of cyber attacks, businesses need effective risk management strategies. Game theory enables organizations and cybersecurity experts to analyze potential attack scenarios, implement preventive measures, and design security systems considering the strategies of both attackers and defenders.

4. **Climate Change Adaptation:** Game theory assists governments and policymakers in making decisions related to climate change adaptation. It helps explore strategies for allocating resources, designing incentives for risk reduction, and coordinating international efforts to tackle common environmental challenges.

Challenges and Limitations in Game Theory Applications

While game theory offers valuable insights into risk management and insurance, it also faces some challenges and limitations:

- **Simplifying Assumptions:** Game-theoretic models often rely on certain assumptions about players' rationality, information availability, and strategies. Real-world situations may deviate from these assumptions, affecting the accuracy of predictions and optimal strategies derived from the models.

- **Complexity and Computational Challenges:** As the number of players and potential strategies increases, game-theoretic models become more complex and computationally intensive. Solving large-scale games with many players and intricate strategies may require advanced computational techniques.

- **Limited Predictive Power:** While game theory provides insights into strategic decision-making, it does not always predict real-world outcomes accurately. The behavior and decision-making of individuals in complex social and economic systems can be challenging to capture within game-theoretic frameworks.

Conclusion

Game theory offers a powerful framework to analyze risk management and insurance problems, providing insights into decision-making, strategies, and optimal risk-sharing mechanisms. By understanding the interactions and strategic behavior of stakeholders, we can design effective risk management strategies, develop fair insurance contracts, and tackle complex challenges in various industries. While game theory has its limitations, its application in risk management and insurance continues to evolve, empowering organizations and individuals to make informed decisions in the face of uncertainty.

Financial Market Regulation and Game Theory

Financial markets play a crucial role in the global economy, facilitating the exchange of financial assets and determining the allocation of capital. However, these markets are not immune to imperfections and failures. To ensure the stability and efficiency of financial markets, regulatory authorities implement various measures that are often influenced by game theory. In this section, we will explore the relationship between financial market regulation and game theory, examining how game-theoretic strategies can guide regulators in addressing market failures and promoting fair and transparent financial systems.

The Need for Financial Market Regulation

Financial markets are subject to several inherent problems, such as information asymmetry, moral hazard, and adverse selection. These issues can undermine the integrity and efficiency of markets, leading to market failures and systemic risks. Without proper regulation, participants in financial markets may engage in unethical behavior, manipulate prices, or take excessive risks, jeopardizing the well-being of individual investors and the overall economy.

The primary objective of financial market regulation is to promote stability, fairness, and transparency. Regulators aim to prevent fraud and market manipulation, ensure the disclosure of accurate and relevant information, and protect investors from unfair practices. Game theory provides a valuable toolset for regulators to analyze and address the strategic interactions among market participants, enabling them to design effective regulatory frameworks and policies.

Principles of Financial Market Regulation

Game theory offers insights into the strategic behavior of market participants, allowing regulators to anticipate the possible responses to regulation and design effective policies. The following principles guide the application of game theory in financial market regulation:

1. **Incentive Alignment**: Regulators must align the incentives of market participants with the desired outcomes. By understanding the motives and strategic behaviors of various participants, regulators can design incentives that encourage responsible behavior and discourage misconduct.

2. **Information Disclosure**: Overcoming information asymmetry is a critical aspect of financial market regulation. Regulators can use game theory to analyze the impact of information disclosure requirements on market outcomes. By imposing disclosure rules, regulators ensure that relevant information is made available to all market participants, reducing information asymmetry and fostering market efficiency.

3. **Monitoring and Enforcement**: Regulators need to monitor the activities of market participants and enforce compliance with regulations. Game theory helps regulators develop monitoring strategies that incentivize compliance and detect potential violations. Moreover, by understanding the strategic interactions between regulators and market participants, regulators can design enforcement mechanisms that mitigate strategic behavior and deter misconduct.

4. **Market Structure Design**: Regulators have the authority to shape the structure of financial markets. Game theory provides insights

into how market structure affects the behavior of participants and market outcomes. By designing market structures that discourage manipulative practices and promote competition, regulators can enhance market efficiency and integrity.

5. **Systemic Risk Management**: Financial market regulation aims to mitigate systemic risks that can threaten the stability of the entire financial system. Game theory helps regulators identify and assess the strategic interactions that contribute to systemic risk. By understanding the interdependencies and feedback effects among market participants, regulators can develop measures to contain systemic risks and prevent financial crises.

Game-Theoretic Tools in Financial Market Regulation

Regulators employ various game-theoretic tools to address specific challenges in financial market regulation. Here are some notable examples:

1. **Market Manipulation Detection**: Game theory can help regulators develop algorithms and models to detect market manipulation, such as spoofing or front-running. By analyzing patterns of trading behavior and strategic interactions, regulators can identify suspicious trading activities and take appropriate enforcement actions.

2. **Optimal Penalty Design**: Game theory can guide regulators in determining the optimal level of penalties for market misconduct. By considering the strategic responses of potential wrongdoers, regulators can set penalties that sufficiently deter misconduct while avoiding excessive punishment that may discourage legitimate market activities.

3. **Risk-Based Supervision**: Game theory assists regulators in designing risk-based supervision frameworks. By evaluating the strategic interactions among financial institutions, regulators can identify the sources of systemic risk and allocate regulatory resources accordingly. This approach enables regulators to focus on institutions that pose a higher risk to the financial system.

4. **Stress Testing**: Game theory plays a crucial role in stress testing methodologies. By modeling the strategic behaviors of market participants under stressful market conditions, regulators can assess the resilience of financial institutions and the stability of the overall financial system. This analysis helps regulators develop effective risk management and contingency plans.

Real-World Examples

To illustrate the application of game theory in financial market regulation, let's consider a couple of real-world examples:

Insider Trading Regulation: Insider trading refers to the practice of trading securities based on non-public information. Regulators combat insider trading by enforcing strict regulations and imposing penalties. Game theory helps regulators understand the strategic incentives of insiders and design regulations that deter insider trading. For example, by increasing the probability and severity of punishment while providing incentives for whistleblowers, regulators can create a stronger deterrent against this unethical behavior.

Capital Adequacy Regulation: Capital adequacy regulation aims to ensure that financial institutions maintain sufficient capital to absorb unexpected losses. Regulators use game theory to analyze the strategic interactions between regulators and banks. By imposing minimum capital requirements and conducting stress tests, regulators can incentivize banks to maintain adequate capital buffers, reducing the likelihood of bank failures and systemic risks.

Key Takeaways

Financial market regulation and game theory are intertwined, with game-theoretic strategies providing valuable insights for regulators in addressing market failures, promoting stability, fairness, and transparency. By understanding the strategic behaviors of market participants, regulators can design effective regulatory frameworks, shape market structures, and enforce compliance. The application of game theory in financial market regulation helps maintain the integrity of financial markets and safeguard the overall stability of the economy.

Exercises

1. Discuss the role of information disclosure in financial market regulation. Provide examples of regulations that aim to reduce information asymmetry and promote market efficiency.

2. Consider a scenario where two financial institutions engage in collusion to manipulate stock prices. Apply game theory to analyze the strategic interactions between the regulators and the colluding institutions. How can regulators detect and deter such collusive behaviors?

3. Research a recent case involving financial market misconduct, such as market manipulation or insider trading. Describe the regulatory actions taken and discuss how game-theoretic tools could have influenced the regulatory response.

Additional Resources

1. Dixit, A., & Pindyck, R. (1994). *Investment Under Uncertainty*. Princeton University Press.

2. Tirole, J. (2017). *Economics for the Common Good*. Princeton University Press.

3. Bolton, P., & Dewatripont, M. (2005). *Contract Theory*. MIT Press.

4. Heath, T., Jackson, M., & Oprea, R. (2015). *Handbook of Financial Intermediation and Banking*. North-Holland.

Remember, the world of financial market regulation is constantly evolving, influenced by new developments in technology, market behavior, and regulatory approaches. Stay curious, keep learning, and always strive for a fair and efficient financial system.

Game Theory and Systemic Risk

In this section, we will explore how game theory can help us understand and manage systemic risk in financial markets. Systemic risk refers to the risk of a widespread financial collapse or disruption that affects the entire financial system, surpassing the risk associated with individual assets or institutions.

Understanding Systemic Risk

Systemic risk arises due to the interconnectedness and interdependencies of financial institutions, markets, and their participants. When one institution or market experiences distress or failure, it can quickly spread to other entities and markets, potentially leading to a cascading effect. This can result in a significant decline in the overall stability and functioning of the financial system.

To understand systemic risk, we need to analyze the interactions and strategic decisions made by financial institutions and market participants. This is where game theory comes into play. Game theory provides a framework for studying the behavior and strategies of multiple players in situations of conflict or cooperation. By applying game theory to systemic risk analysis, we can gain insights into the dynamics and vulnerabilities of the financial system.

Game Theory Models for Systemic Risk

In the context of systemic risk, game theory can be used to model the strategic interactions and decision-making processes of financial institutions. One commonly used game-theoretic model is the "Bank Run" game, which captures the contagious nature of financial distress. In this game, multiple banks face the decision of whether to withdraw their funds from a troubled bank or continue to hold their deposits. The outcome of the game depends on the expectations and actions of each bank.

Another important game-theoretic model for systemic risk is the "Coordination Game." This model examines situations where financial institutions need to coordinate their actions to prevent or mitigate a crisis. For example, during a market panic, if all institutions simultaneously sell their assets, it can worsen the crisis. However, if they coordinate and collectively agree not to sell, they can stabilize the market. The challenge lies in reaching a cooperative equilibrium where all participants trust each other's actions and refrain from panicking.

Strategies to Manage Systemic Risk

Game theory provides insights into strategies that can help manage systemic risk and enhance financial stability. Here are some key strategies:

1. **Early Warning Systems**: By employing game-theoretic models, regulators can develop early warning systems to identify and monitor potential sources of systemic risk. These systems rely on analyzing market data, participants' behavior, and network connections to detect emerging risks and take appropriate preventive measures.

2. **Macroprudential Policies**: Game theory can guide policymakers in designing and implementing macroprudential policies to reduce systemic risk. These policies aim to enhance the resilience of the financial system as a whole. For example, setting capital requirements, implementing stress tests, and establishing risk-sharing mechanisms can help mitigate the effects of financial shocks.

3. **Central Bank Interventions**: Central banks play a crucial role in managing systemic risk by acting as the lender of last resort. Game theory can inform the decision-making process of central banks during crises, such as providing liquidity support and implementing measures to restore confidence and stability in financial markets.

4. **Enhancing Cooperation**: Cooperation among financial institutions is essential for managing systemic risk effectively. By analyzing strategic interactions using game theory, policymakers can design incentives and mechanisms that encourage cooperation and discourage behaviors that amplify systemic risk. For instance, the establishment of clearinghouses and information-sharing arrangements can facilitate coordination and reduce the potential for contagion.

Real-world Applications

Game theory has found practical applications in managing systemic risk in various real-world scenarios. One notable example is the establishment of the Committee on Payments and Market Infrastructures (CPMI) and the International Organization of Securities Commissions (IOSCO) Principles for Financial Market Infrastructures. These principles aim to enhance the resilience of payment and settlement systems, reducing the risk of widespread disruptions.

Another application is the adoption of game-theoretic models in stress testing and scenario analysis by regulatory authorities. These models help assess the vulnerability of the financial system to adverse shocks and inform policymakers about potential systemic risks and their implications.

Summary

Systemic risk poses a significant threat to the stability and functioning of financial markets. By applying game theory, we can analyze the strategic interactions and decision-making processes of financial institutions and develop strategies to mitigate systemic risk. Early warning systems, macroprudential policies, central bank interventions, and enhancing cooperation are crucial elements in managing systemic risk effectively. Real-world applications have demonstrated the value of game theory in identifying vulnerabilities and improving the resilience of the financial system.

In the next section, we will explore the role of game theory in the field of artificial intelligence and its implications for strategic interaction and decision-making.

Real-World Examples of Game Theory in Financial Markets

In this section, we will explore several real-world examples that demonstrate the application of game theory in financial markets. Game theory provides a valuable framework for understanding the strategic behavior of market participants and the dynamics of market interactions. By modeling the interaction between buyers and sellers, investors and financial institutions, and regulators and market participants, game theory offers insights into the complexities of financial markets and helps us make better-informed decisions.

Example 1: The Prisoner's Dilemma in the Stock Market

The Prisoner's Dilemma is a classic game theory scenario that has implications in the stock market. Imagine two traders, Alice and Bob, who are both deciding whether to buy or sell a particular stock. They have the option to either cooperate (buy) or defect (sell). If both traders cooperate, they can collectively drive up the stock price and make a profit. However, if one trader defects while the other cooperates, the defector can take advantage of the cooperative trader's action and make an even larger profit. If both traders defect, they may drive down the stock price and both end up with losses.

In this scenario, each trader must weigh the potential risks and benefits of their actions. If Alice believes that Bob will cooperate, it would be in her best interest to defect and take advantage of the situation. However, if Alice believes that Bob will defect, she should also defect to minimize her losses. Bob faces the same decision-making process.

This example illustrates how game theory can be used to analyze the strategic interactions in the stock market. It highlights the importance of considering the actions and expectations of other market participants when making investment decisions. By understanding the incentives and motivations of market players, investors can better assess the potential risks and rewards associated with their investment choices.

Example 2: Oligopoly Pricing and Collusion

Oligopoly refers to a market structure in which a small number of firms dominate the industry. In such markets, game theory plays a crucial role in understanding pricing strategies and the potential for collusive behavior among firms.

Consider the airline industry, where a few major carriers control a significant portion of the market. These airlines must make strategic decisions regarding ticket pricing, taking into account their competitors' actions. If one airline lowers its prices, it may attract more customers, but at the expense of lower profitability. Other airlines have to respond strategically to either match the lower prices or maintain higher prices to protect their profit margins.

Game theory helps us understand the pricing dynamics in this oligopoly market. Pricing decisions are akin to a game, where each firm must carefully analyze the potential reactions of its competitors. If these firms were to collude and agree on a high pricing strategy, they could collectively maximize their profits. However, such collusion is often illegal and can be difficult to maintain, leading to intense competition and the possibility of price wars.

The study of game theory in oligopolistic markets allows regulators and policymakers to assess the effectiveness of antitrust laws and regulations. It helps predict price changes, market behaviors, and the implications of different market structures on consumer welfare.

Example 3: Algorithmic Trading and Market Manipulation

In modern financial markets, algorithmic trading has become increasingly prevalent. Algorithmic trading refers to the use of computer programs to execute trades in financial markets, often based on predefined rules and strategies. Game theory provides a valuable framework for understanding the strategic interactions between algorithmic traders and traditional market participants.

One example of how game theory applies to algorithmic trading is through the concept of market manipulation. Traders can use sophisticated algorithms to manipulate the market by inducing others to trade based on false or misleading information. By strategically placing orders and canceling them, these traders can create artificial price movements to their advantage.

Regulators and market participants need to understand these manipulative strategies and develop countermeasures based on game theory. By modeling the behavior of algorithmic traders and identifying their incentives, regulators can design regulations that discourage manipulative practices and ensure fair and transparent markets.

Overall, game theory provides a powerful tool to analyze and understand the complexities of financial markets. By incorporating game-theoretic principles into financial decision-making, market participants can better navigate the strategic interactions, anticipate the actions of other players, and ultimately make more informed investment choices. Game theory continues to evolve alongside the ever-changing landscape of financial markets, playing a crucial role in shaping the future of finance.

Game Theory in Artificial Intelligence

Machine Learning and Game Theory

Reinforcement Learning and Game Theory

In this section, we will explore the fascinating intersection of reinforcement learning and game theory. Reinforcement learning is a subfield of artificial intelligence that focuses on how intelligent agents can learn optimal behaviors through interaction with their environment. Game theory, on the other hand, studies strategic decision-making in situations where the outcome of one player's decision depends on the decisions of others. By combining these two fields, we can delve into the dynamics of decision-making in multi-agent environments and uncover strategies that lead to optimal outcomes.

Reinforcement Learning: A Brief Overview

Before we dive into the connection between reinforcement learning and game theory, let's first understand the fundamentals of reinforcement learning itself. At its core, reinforcement learning is based on the concept of an agent, an entity that interacts with an environment and learns from its actions. The agent's goal is to maximize some notion of cumulative reward over time by selecting actions based on the environment's state.

The fundamental components of a reinforcement learning problem are as follows:

1. **State:** A state represents the current configuration of the environment, which the agent perceives.

2. **Action:** An action is a decision taken by the agent at a particular state.

3. **Reward:** A reward is a scalar value that provides feedback to the agent about the quality of its action at a specific state. The agent's objective is to maximize the cumulative reward over time.

4. **Policy:** A policy defines the agent's strategy for selecting actions based on the observed state.

5. **Value function:** A value function estimates the expected cumulative reward that an agent can achieve from a given state or action.

Reinforcement learning can be viewed as a Markov decision process (MDP), which is a mathematical framework for modeling decision-making problems. MDPs capture the properties of sequential decision-making by assuming a probabilistic transition between states and a reward signal associated with each state-action pair.

Solving an MDP involves finding an optimal policy that maximizes the expected cumulative reward. This can be achieved by using various algorithms, such as value iteration, policy iteration, or Monte Carlo methods, to estimate the value function associated with different states or state-action pairs.

Game Theory: A Brief Overview

Now that we have a good understanding of reinforcement learning, let's explore the key concepts of game theory. Game theory provides a framework for analyzing strategic interactions between rational decision-makers, known as players, who aim to maximize their individual outcomes.

A game in the context of game theory consists of the following components:

1. **Players:** Players are the decision-makers or entities involved in the game. Each player selects an action or strategy based on their preferences and the actions of other players.

2. **Strategies**: Strategies represent the set of actions available to each player.

3. **Payoffs**: Payoffs quantify the outcome or utility that each player receives based on the combination of strategies chosen by all players.

4. **Nash Equilibrium**: Nash equilibrium is a concept in game theory that represents a stable state in which no player has an incentive to unilaterally change their strategy. It is a solution concept used to predict the outcome of a game when players are rational.

Game theorists seek to identify Nash equilibria and understand the underlying strategies that lead to optimal outcomes in various game scenarios, such as the Prisoner's Dilemma, Battle of the Sexes, or Chicken game.

Reinforcement Learning and Game Theory: An Integration

The integration of reinforcement learning and game theory allows us to study the dynamics of decision-making in multi-agent environments. In this setting, multiple agents interact with each other, and their collective actions lead to outcomes that impact individual rewards.

Reinforcement learning algorithms can be extended to address multi-agent scenarios by introducing game-theoretic concepts. One such approach is known as multi-agent reinforcement learning (MARL), where multiple agents learn to coordinate their actions and optimize their own rewards by considering the actions of others.

MARL algorithms can be classified into two main categories: independent learner models and joint action learner models. In independent learner models, each agent learns independently by treating other agents as part of the environment. On the other hand, joint action learner models explicitly model the interactions between agents and aim to find optimal joint policies that maximize the collective reward.

One notable algorithm in the realm of MARL is the Q-learning algorithm with Nash Q-learning, which extends the classical Q-learning algorithm to handle adversarial interactions between agents. By computing Nash equilibria during the learning process, agents can converge to strategies that are not exploitable by others, leading to more robust decision-making and optimal outcomes.

Applications and Challenges

Reinforcement learning combined with game theory has a wide range of applications in various domains. Some notable examples include:

- **Robotics**: Utilizing game-theoretic reinforcement learning algorithms, robots can learn to collaborate and coordinate their actions to accomplish complex tasks, even in uncertain or adversarial environments.

- **Economics and Market Dynamics**: Game-theoretic analysis combined with reinforcement learning can shed light on the emergence of market equilibria, price dynamics, and strategic interactions between market participants.

- **Traffic Management**: By applying game-theoretic reinforcement learning, traffic signal timings can be optimized to minimize congestion and maximize traffic flow in urban areas.

However, the integration of reinforcement learning and game theory also presents several challenges. One of the main challenges is finding computationally tractable algorithms that can effectively learn and converge to optimal strategies in complex, multi-agent settings. Additionally, the presence of multiple agents with potentially conflicting objectives introduces the issue of fairness and the need to design mechanisms that promote cooperation and mitigate selfish behaviors.

Real-World Example: Cooperative Multi-Agent Systems

Let's consider a real-world example of cooperative multi-agent systems, where a group of autonomous drones needs to coordinate their actions to accomplish tasks efficiently.

In this scenario, each drone acts as an independent learner and aims to maximize its own reward by performing the assigned task. However, there is interdependence among the drones because the overall team performance depends on how well they collaborate.

By applying reinforcement learning algorithms with game-theoretic principles, the drones can learn optimal strategies that balance both individual and collective rewards. Nash equilibria can be computed during the learning process to ensure that no drone has an incentive to deviate from its strategy.

Through repeated interactions and learning, the drones can acquire efficient coordination strategies, leading to improved task completion times, reduced energy consumption, and enhanced overall team performance.

Conclusion

The integration of reinforcement learning and game theory provides a powerful framework for understanding strategic decision-making in multi-agent scenarios. By leveraging the principles of reinforcement learning and game theory, we can analyze complex interactions, design optimal strategies, and address real-world challenges in various domains.

From robotics and economics to traffic management and beyond, the combination of reinforcement learning and game theory opens up new avenues for exploration and innovation. As we delve deeper into this field, we will continue to unravel the intricacies of decision-making and unlock the potential of intelligent, strategic agents.

Game Theory in Multi-Agent Systems

In this section, we will explore the application of game theory in the context of multi-agent systems (MAS). A multi-agent system refers to a collection of autonomous agents that interact with each other to achieve individual or collective goals. Game theory provides a powerful framework to analyze and understand the strategic interactions between these agents in various domains such as economics, computer science, and social networks.

Basic Concepts of Multi-Agent Systems

Before delving into the application of game theory, let's briefly discuss some fundamental concepts of multi-agent systems.

In a multi-agent system, each agent is a decision-making entity with its own goals, knowledge, and capabilities. These agents can perceive the environment, take actions, and interact with other agents. The interactions can be cooperative, competitive, or a mix of both.

The main elements of a multi-agent system are:

- **Agents:** Autonomous entities that make decisions and interact with each other.

- **Environment:** The context in which the agents operate.

- **Actions:** The choices that the agents can make.

- **Utility or Payoff:** The measure of the desirability or satisfaction for an agent associated with the outcomes of its actions.

Strategic Interactions in Multi-Agent Systems

In multi-agent systems, agents often make decisions based on their strategic interactions with other agents. These interactions can be modeled using game theory, which provides a mathematical framework to study strategic decision-making.

Game theory analyzes the interactions among rational agents who aim to maximize their own utility or payoff. It provides tools and concepts to model, analyze, and predict the behavior of agents in various scenarios.

Game Theory in Multi-Agent Systems

Now let's explore how game theory can be applied to study and analyze multi-agent systems.

Nash Equilibrium: One of the central concepts in game theory is the Nash equilibrium. In a multi-agent system, a Nash equilibrium is a state in which no agent has an incentive to unilaterally deviate from its chosen strategy, given the strategies chosen by other agents. It represents a stable outcome where no agent can improve its payoff by changing its strategy. Nash equilibrium serves as a powerful tool to predict the behavior of agents in different scenarios.

Coordination Games: Coordination games are one type of game often encountered in multi-agent systems. In these games, agents need to coordinate and align their actions to achieve a common goal. A famous example is the Battle of the Sexes game, where a couple needs to decide on a mutually convenient activity. Game theory provides strategies and analysis to ensure efficient coordination among agents.

Prisoner's Dilemma: The Prisoner's Dilemma is another classical game in game theory that finds applications in multi-agent systems. In this game, two individuals are arrested for a crime and face the dilemma of whether to cooperate with each other or betray their partner. The dilemma arises from the conflict between individual and collective rationality. The Prisoner's Dilemma helps understand the challenges of cooperation in multi-agent systems and provides insights into strategies that promote cooperation despite self-interest.

Evolutionary Game Theory: In a dynamic multi-agent system, where agents can adapt and learn from their interactions, evolutionary game theory offers valuable insights. It combines game theory with evolutionary concepts to study the emergence and evolution of strategies over time. By introducing mechanisms of mutation, reproduction, and natural selection, evolutionary game theory helps analyze how strategies spread and dominate in populations of agents.

Applications of Game Theory in Multi-Agent Systems

Game theory has a wide range of applications in multi-agent systems. Let's explore a few examples:

Robotics: In the field of robotics, game theory can be used to analyze the interactions and decision-making of multiple robots. By modeling their behavior as a game, researchers can optimize task allocation, resource sharing, and coordination strategies to improve overall system performance.

Internet Routing: In the domain of computer networks, game theory can help us understand and optimize the routing decisions made by routers in the internet infrastructure. By analyzing the interactions between routers as a game, researchers can design efficient routing protocols that minimize congestion and ensure the smooth flow of data.

Smart Grids: In the context of energy systems, game theory can aid in the design of pricing mechanisms and energy trading strategies among multiple participants. By modeling the interactions between energy producers, consumers, and grid operators as a game, researchers can design strategies that incentivize renewable energy generation and promote the stability of the grid.

Market Competition: Game theory is well-established in the study of economics, where it helps analyze and predict the behavior of firms in competitive markets. By modeling the interactions between firms as a game, researchers can study strategies such as pricing, product differentiation, and market entry/exit decisions.

Conclusion

In conclusion, game theory provides valuable tools and concepts to analyze strategic interactions and decision-making in multi-agent systems. By applying game theory, we can understand and predict the behavior of agents in various domains, ranging from robotics to economics. Game theory offers insights into optimal strategies, the emergence of cooperation, and the dynamics of decision-making in complex environments. Understanding game theory in the context of multi-agent systems is crucial for designing intelligent and adaptive systems in the modern world.

Practice Problems

1. Consider a multi-agent system where two competing companies are deciding whether to invest in research and development (R&D) or marketing. The payoffs associated with each strategy are as follows:
    a) Is there a Nash equilibrium in this game? If so, what is it?

Company 1/Company 2	Invest in R&D	Invest in Marketing
Invest in R&D	3, 3	0, 4
Invest in Marketing	4, 0	2, 2

b) What would be the best strategy for each company to maximize its payoff?

2. Consider a scenario where two teams of soccer players are competing against each other. Each player has to choose between an offensive strategy and a defensive strategy. The payoffs associated with each strategy are as follows:

Team 1/Team 2	Offensive Strategy	Defensive Strategy
Offensive Strategy	4, 2	1, 3
Defensive Strategy	3, 1	2, 0

a) Is there a Nash equilibrium in this game? If so, what is it?

b) What would be the best strategy for each team to maximize its payoff?

3. Think of a real-world multi-agent system where game theory can be applied. Describe the agents, their interactions, and the potential outcomes that game theory can help analyze.

Resources and Further Reading

- Osborne, M. J., and Rubinstein, A. (1994). *A Course in Game Theory*. MIT Press.

- Leyton-Brown, K., and Shoham, Y. (2008). *Essentials of Game Theory: A Concise, Multidisciplinary Introduction*. Morgan & Claypool Publishers.

- Björnerstedt, J., and Holmberg, S. (2020). *Game Theory for Business and Management*. Cambridge University Press.

- Gintis, H. (2009). *Game Theory Evolving: A Problem-Centered Introduction to Modeling Strategic Behavior*. Princeton University Press.

- Shoham, Y., and Leyton-Brown, K. (2009). *Multiagent Systems: Algorithmic, Game-Theoretic, and Logical Foundations*. Cambridge University Press.

Note: For solutions to the practice problems and additional resources, please refer to the Appendix section at the end of the book.

Now take a break, stretch your legs, and get ready for the next section where we will explore the fascinating relationship between game theory and social networks!

Game-Theoretic Approaches to AI

In recent years, game theory has emerged as a powerful tool in the field of artificial intelligence (AI). By leveraging the principles of strategic decision-making and analyzing the interaction between multiple intelligent agents, game-theoretic approaches have revolutionized the way AI systems are designed and developed. This section explores the various applications of game theory in AI, highlighting its role in understanding and modeling complex interactions, decision-making, and learning.

Strategic Decision-Making in AI

Game theory provides AI systems with a framework for strategic decision-making in multi-agent environments. In such environments, an AI agent's actions can have a direct impact on the strategies and outcomes of other agents. By understanding the strategic choices available to other agents and anticipating their responses, AI agents can optimize their decision-making process.

One common game-theoretic approach used in AI is the concept of Nash equilibrium. A Nash equilibrium is a set of strategies where no agent can unilaterally improve its outcome by deviating from its strategy while other agents keep their strategies unchanged. AI systems can use algorithms to compute Nash equilibria in games, allowing them to make rational and optimal decisions based on the expected actions of other agents.

Learning Strategies from Game Theory

Game-theoretic approaches also empower AI systems to learn optimal strategies based on past experiences and interactions. Through a process known as reinforcement learning, AI agents can adapt and improve their behaviors by receiving feedback and rewards. In game-theoretic reinforcement learning, agents learn to optimize their strategies by observing the actions and outcomes of other agents.

One popular algorithm used in game-theoretic reinforcement learning is called fictitious play. In this algorithm, an AI agent assumes that its opponents are playing with fixed strategies and updates its own strategy based on the observed frequencies of opponent moves. By continuously updating and adapting its strategy, the AI agent can converge towards optimal play, even in dynamic and uncertain environments.

Cooperative and Competitive AI

Game theory enables AI systems to model both cooperative and competitive interactions. In cooperative games, AI agents work together towards a common goal, sharing resources and coordinating their actions. By applying cooperative game theory, AI agents can allocate resources fairly, determine the value of contributions from each agent, and ensure stable cooperation.

In contrast, competitive games involve conflicting interests and limited resources. In such scenarios, AI agents aim to maximize their own individual goals while anticipating and countering the strategies of other agents. By employing competitive game theory, AI agents can develop strategies that exploit weaknesses in opponents' behaviors and make rational decisions to gain an advantage.

Game-Theoretic Models in AI Research

Game-theoretic approaches have found applications in various domains within AI research. One such area is autonomous vehicle navigation. When multiple autonomous vehicles share the same road network, they need to make decisions about speed, lane changes, and interactions with other vehicles. Using game-theoretic models, AI agents can make decisions that balance efficiency, safety, and cooperation among vehicles.

Another application is in algorithmic trading, where AI agents compete to maximize their profits in financial markets. By incorporating concepts from game theory, AI agents can analyze the behaviors of other market participants, predict price movements, and make informed trading decisions in a dynamic and competitive environment.

Ethical Implications and Limitations

While game theory offers valuable insights and tools for AI, it also raises important ethical considerations. For instance, in competitive settings, AI agents may exploit vulnerabilities or engage in deceptive strategies to gain an advantage. Balancing the pursuit of optimal outcomes with ethical considerations is crucial to ensure responsible AI development.

Furthermore, game-theoretic approaches often assume rationality and perfect information, which may not hold in real-world scenarios. The complexity of modeling and analyzing interactions between multiple intelligent agents can also lead to computational challenges and scalability issues.

Conclusion

In conclusion, game-theoretic approaches play a vital role in AI by enabling strategic decision-making, learning optimal strategies, and modeling cooperative and competitive interactions. With the power to anticipate and respond to others' actions, AI systems can navigate complex environments, optimize resource allocation, and make rational choices. Despite ethical considerations and computational limitations, game theory continues to shape the future of AI, driving advancements in various domains and revolutionizing the way intelligent agents interact and learn.

Algorithmic Game Theory

In the previous chapters, we explored various aspects of game theory, from decision-making and types of games to social networks and evolutionary biology. Now, let's dive into the fascinating world of algorithmic game theory, where we explore the interplay between algorithms and strategic interactions.

Understanding Algorithmic Game Theory

Algorithmic game theory combines computer science and economics to study strategic decision-making in computational settings. It focuses on designing algorithms and computational models that capture the complexities of strategic interactions, analyzing their properties, and understanding their implications.

In algorithmic game theory, we consider a game as a computational problem with players, strategies, and payoffs. The goal is to develop algorithms that can efficiently find optimal solutions, determine equilibria, and analyze the convergence of game dynamics.

Computational Complexity and Game Theory

One of the central questions in algorithmic game theory is understanding the computational complexity of finding equilibria in games. The complexity class NP (nondeterministic polynomial time) plays a crucial role in this analysis.

The class NP consists of decision problems that can be verified in polynomial time. For example, given a solution to a problem, we can efficiently verify whether it is correct. However, finding these solutions efficiently remains an open question.

In game theory, the class PPAD (Polynomial Parity Argument, Directed version) is particularly relevant. It contains problems that capture the computational complexity of finding pure Nash equilibria, a key concept in game theory. Determining whether PPAD is equal to NP is a fundamental open problem in algorithmic game theory.

Algorithmic Mechanism Design

Algorithmic mechanism design is the branch of algorithmic game theory concerned with designing incentive-compatible mechanisms for achieving desired outcomes in strategic settings. It aims to construct algorithms that incentivize strategic players to behave in a way that aligns with societal objectives.

One of the foundational concepts in algorithmic mechanism design is the revelation principle. According to this principle, for any mechanism, there exists an equivalent one in which players truthfully reveal their preferences. This simplifies the design of mechanisms by allowing us to focus on truthful mechanisms.

A well-known mechanism design problem is combinatorial auctions, where multiple goods are sold simultaneously to bidders with complex preferences. Designing efficient algorithms for combinatorial auctions involves dealing with computational challenges like winner determination and allocation of goods.

Game Dynamics and Learning

Game dynamics refer to the process by which a game evolves over time as players interact and make decisions. Strategic interactions often involve repeated games, where players' past actions influence their future decisions. Understanding how these dynamics play out is essential in algorithmic game theory.

Reinforcement learning is a popular framework from machine learning that addresses how agents can learn to make optimal decisions in dynamic environments. In the context of algorithmic game theory, researchers explore how reinforcement learning algorithms can converge to equilibria or achieve desirable social outcomes.

Learning in games is often analyzed through the lens of regret. Regret measures the deviation of a player's strategy from that of the optimal strategy in hindsight. Exploring algorithms that achieve low regret and converge to equilibrium is a central research topic in algorithmic game theory.

Game Theory in Online Advertising

Let's take a real-world example to illustrate the application of algorithmic game theory. In online advertising, advertisers compete for ad placements. Each advertiser wants to maximize its own utility, which may depend on various factors such as click-through rates, conversion rates, and budget constraints.

The allocation of ad placements and pricing can be modeled as an auction game. Advertisers bid for ad slots, and an algorithm determines the allocation of slots and the prices they pay. This process involves algorithmic game theory techniques such as mechanism design, revenue maximization, and optimizing for efficiency.

The challenge lies in designing efficient algorithms that balance the interests of advertisers, the ad platform, and the users. Algorithmic game theory enables us to model the interactions between these stakeholders and develop mechanisms that achieve desirable outcomes, such as maximizing overall revenue or user satisfaction.

Summary

In this section, we explored algorithmic game theory, a thrilling field that combines computer science and economics. We discussed the interplay between algorithms and strategic interactions, delving into computational complexity, algorithmic mechanism design, game dynamics, and learning.

Algorithmic game theory provides powerful tools for understanding and analyzing strategic interactions in various domains, from online advertising to resource allocation and market design. By leveraging computational models and algorithmic approaches, we can design mechanisms that align the interests of different stakeholders and achieve desirable outcomes.

As technology continues to advance and shape our society, algorithmic game theory will play an increasingly crucial role in addressing complex strategic problems and designing efficient solutions. Whether it's optimizing auctions, developing fair resource allocation algorithms, or understanding the dynamics of social networks, algorithmic game theory offers fascinating insights and practical applications.

Game Theory and Autonomous Systems

As technology advances at an exponential rate, autonomous systems are becoming increasingly prevalent in our everyday lives. From self-driving cars to robotic assistants, these intelligent machines are capable of making decisions and taking actions on their own. But how do we ensure that these autonomous systems behave in a way that aligns with our goals and values? This is where the intersection of game theory and autonomous systems comes into play.

Understanding Autonomous Systems

Before diving into the game theory aspects, let's first understand what autonomous systems are. An autonomous system is a machine or software that can perform tasks or make decisions without direct human intervention. These systems rely on sensors, algorithms, and machine learning to gather information, analyze it, and act accordingly.

Autonomous systems can possess a wide range of capabilities, from simple decision-making in controlled environments to complex reasoning and problem-solving in dynamic and uncertain situations. They are designed to optimize certain objectives, such as efficiency, safety, or cost-effectiveness, while adhering to predefined rules or constraints.

Game Theory and Rational Decision-Making

Game theory provides a framework for analyzing strategic interactions among multiple agents. In the context of autonomous systems, game theory helps us understand how these systems make decisions when faced with uncertainty and the presence of other agents. By incorporating game-theoretic principles, we can ensure that autonomous systems act rationally and effectively in different scenarios.

At the heart of game theory lies the concept of a game. A game comprises players, strategies, and payoffs. In the case of autonomous systems, the players may include other autonomous agents, humans, or a combination of both. Strategies represent the different actions available to the system, while payoffs quantify the benefits or costs associated with each outcome.

By employing game theory, we can model the decision-making process of autonomous systems by considering the knowledge, preferences, and goals of the system. This allows us to analyze and predict the behavior of the system in different scenarios, improving our understanding of its actions and their consequences.

Cooperation and Competition in Autonomous Systems

Autonomous systems often operate in environments where cooperation and competition coexist. Take, for example, a swarm of autonomous drones tasked with delivering packages in a densely populated area. These drones need to cooperate to ensure efficient delivery routes, yet they also compete for resources such as airspace and landing spots.

Game theory provides valuable insights into the cooperation and competition dynamics between autonomous systems. It can help identify strategies that maximize collective benefits, minimize conflicts, and establish stable cooperative or competitive equilibria.

For instance, in a scenario where autonomous vehicles share the road, game theory can be used to study the behavior of these vehicles and identify optimal strategies that minimize traffic congestion, reduce accidents, and improve overall transportation efficiency. By strategically choosing routes, speeds, and actions, autonomous vehicles can navigate through complex traffic scenarios while optimizing for the collective benefit of all vehicles on the road.

Game Theory and Decision Fusion

In many situations, autonomous systems need to make group decisions based on information from multiple sources. This is known as decision fusion, and it poses unique challenges for autonomous systems.

Game theory offers a framework to model decision fusion scenarios and optimize the decision-making process. By considering the objectives and capabilities of each autonomous system, game theory allows us to determine how the decisions of individual agents can be merged to arrive at a collective decision that maximizes the overall system performance.

For example, imagine a team of search and rescue drones deployed to locate survivors in a disaster-stricken area. Each drone has limited sensing capabilities but can communicate and collaborate with other drones. By applying game theory, the drones can coordinate their search patterns and fusion algorithms to efficiently cover the area and identify survivors in the shortest possible time.

Ethical Considerations and Autonomous Systems

As autonomous systems become more complex and capable, ethical considerations arise. How do we ensure that these systems adhere to ethical principles and make decisions that align with human values?

Game theory provides a framework to incorporate ethical considerations into the decision-making process of autonomous systems. By defining utilities or payoffs that capture ethical values, we can guide the behavior of these systems in a way that respects societal norms, fairness, and human preferences.

For instance, in the case of autonomous vehicles, game theory can assist in resolving ethical dilemmas such as the famous "trolley problem." By introducing societal preferences and weighting different outcomes, an autonomous vehicle can be programmed to make decisions that prioritize minimizing harm to humans.

Real-World Applications

The integration of game theory into autonomous systems has contributed to various real-world applications. Here are a few examples:

1. Robot-Assisted Surgery: In the field of healthcare, autonomous surgical robots can use game-theoretic models to optimize surgical procedures, taking into account patient-specific factors, the surgeon's expertise, and the desired surgical outcomes.

2. Smart Grid Management: Game theory can enable autonomous systems to cooperate and negotiate in energy management within smart grid networks. This helps balance energy supply and demand, manage peak loads, and encourage the use of renewable energy sources.

3. Swarm Robotics: Game theory plays a crucial role in the coordination and control of swarms of autonomous robots, allowing them to accomplish complex tasks such as mapping, exploration, and disaster response in a decentralized and efficient manner.

4. Cybersecurity: Autonomous systems can leverage game theory to defend against cyber threats by modeling attacker-defender interactions. By considering different attack strategies and defensive measures, autonomous systems can identify optimal security policies and adapt to changing threat landscapes.

Conclusion

Game theory serves as a powerful tool for understanding and improving the decision-making processes of autonomous systems. By incorporating game-theoretic principles, we can design these systems to act rationally, cooperate effectively, and make ethically informed decisions. As autonomous systems continue to evolve and become integral parts of our lives, the intersection of game theory and autonomous systems will continue to shape the future of technology and society as a whole.

Cooperative and Competitive AI

In the world of artificial intelligence (AI), the concepts of cooperation and competition play a vital role. Cooperative AI involves multiple AI agents working together towards a common goal, while competitive AI focuses on AI agents competing against each other to achieve individual objectives. These approaches have significant implications in various domains such as robotics, multi-agent systems, and game playing.

Cooperative AI

Cooperative AI involves the collaboration of AI agents to accomplish a shared objective. The key challenge in cooperative AI is designing strategies for agents to coordinate their actions while maximizing the collective outcome. One popular framework used in cooperative AI is cooperative game theory, which provides mathematical tools to analyze and model cooperative decision-making.

Coalition Formation In cooperative AI, agents often form coalitions to enhance their collective capabilities. The formation of coalitions allows agents to combine their resources, knowledge, and skills, leading to more effective problem-solving. The concept of coalition formation can be modeled using cooperative game theory, where each coalition distributes its payoff among its members.

For example, imagine a group of AI-powered robots working together to clean a large office building. Each robot has a specific set of cleaning tasks it can perform, and by forming coalitions, they can divide the workload efficiently. The challenge lies in determining how to distribute the overall benefit among the robots, taking into account their individual contributions and capabilities.

Coordination Mechanisms Cooperative AI requires effective coordination mechanisms to ensure collaboration among agents. These mechanisms aim to establish communication protocols, shared plans, and decision-making procedures. Without proper coordination, agents may encounter conflicts, redundancies, or inefficiencies in their actions.

One well-known coordination mechanism in cooperative AI is the use of joint-action models. These models define a set of rules and strategies for agents to synchronize their actions in real-time. For example, in a team of autonomous drones delivering packages, coordination mechanisms are essential to avoid mid-air collisions and optimize the delivery routes.

Applications of Cooperative AI Cooperative AI finds applications in various domains, including multi-robot systems, swarm intelligence, and disaster management. In multi-robot systems, cooperative AI enables a group of robots to perform complex tasks collectively, such as search and rescue missions or warehouse automation.

Swarm intelligence, inspired by the collective behavior of social insects, leverages cooperative AI to solve problems involving large numbers of simple agents. Examples include ant colony optimization, where ants collectively find the shortest path between food sources and their nest, or the flocking behavior of birds.

Furthermore, in disaster management scenarios, cooperative AI can assist in coordinating emergency response teams, allocating resources efficiently, and optimizing evacuation plans.

Competitive AI

Competitive AI focuses on the interaction and competition between AI agents. In competitive scenarios, AI agents strive to outperform each other to achieve individual objectives. Game theory, specifically competitive game theory, provides a robust framework for analyzing and designing strategies for competitive AI.

Game-Theoretic Modeling Game-theoretic modeling is an essential tool for competitive AI, allowing the analysis of strategic interactions between AI agents. In competitive game theory, agents aim to maximize their own payoffs while considering the strategies of their opponents. This approach provides insights into decision-making in situations with conflicting interests.

One of the fundamental concepts in competitive game theory is the Nash equilibrium, where no player can unilaterally improve their outcome by changing strategies. Algorithms and techniques are developed to find Nash equilibria, helping AI agents determine optimal strategies in competitive scenarios.

Multi-Agent Adversarial Games Multi-agent adversarial games, such as chess or poker, are classic examples of competitive AI. In these games, AI agents compete directly against each other, employing strategies and tactics to outmaneuver their opponents.

For instance, in chess, each AI agent assesses the board state, considers various moves, and predicts the opponent's responses. The AI agent aims to formulate a strategy that maximizes its chances of winning while predicting and countering the opponent's moves.

Reinforcement Learning in Competitive AI Reinforcement learning is often employed in competitive AI scenarios. The use of reinforcement learning allows AI agents to learn optimal strategies through trial and error. Agents receive rewards or penalties based on their actions, encouraging them to discover effective strategies over time.

For example, in a competitive game like Go, AI agents can use reinforcement learning techniques to improve their moves. They learn from previous experience and adjust their strategies to counter the opponent's moves, leading to more competitive gameplay.

Applications of Competitive AI Competitive AI has numerous applications, including strategic decision-making, financial markets, and cybersecurity. In strategic decision-making, competitive AI models can assist businesses in analyzing market dynamics, predicting competitors' behavior, and optimizing pricing strategies.

In financial markets, AI agents compete to make profitable trades. They analyze market data, identify patterns, and execute trades to maximize profits. The use of competitive AI in financial markets raises several ethical concerns, such as the potential for market manipulation or the impact on market fairness.

Additionally, in the field of cybersecurity, competitive AI plays a crucial role in defending against malicious actors. AI agents compete against attackers in detecting and preventing cyber threats, developing robust intrusion detection systems, and enhancing the security of digital systems.

Summary

Cooperative and competitive AI are two fundamental approaches in artificial intelligence. Cooperative AI focuses on collaboration among AI agents to achieve shared goals, while competitive AI centers around AI agents competing against each other for individual objectives. Both approaches utilize mathematical frameworks, such as game theory, to model and analyze strategic interactions and decision-making.

In the cooperative AI domain, coalition formation and coordination mechanisms are essential for effective collaboration. Cooperative AI finds applications in various fields, including multi-robot systems, swarm intelligence, and disaster management.

On the other hand, competitive AI involves analyzing strategic interactions between AI agents to determine optimal strategies. Game-theoretic modeling and reinforcement learning are valuable tools in competitive AI. Applications of competitive AI range from strategic decision-making and financial markets to cybersecurity.

Understanding and harnessing cooperative and competitive AI is crucial in the development of intelligent systems that can effectively cooperate or compete in a wide range of real-world scenarios. By integrating these approaches thoughtfully, we can unlock the vast potential of AI in addressing complex problems and shaping the future of technology.

Security and Game Theory in AI

In this section, we will explore the fascinating intersection of security and game theory within the realm of artificial intelligence. As AI becomes increasingly prevalent in various domains, it is crucial to understand the security challenges that arise and how game theory can be leveraged to address them. We will delve into the principles, techniques, and real-world applications of security in AI, along with the ethical implications that accompany them.

Security in AI: An Introduction

Security in AI is concerned with safeguarding AI systems and their data from unauthorized access, malicious attacks, and adversarial manipulation. As AI technologies continue to advance, they become susceptible to various security threats, including data breaches, model poisoning, adversarial attacks, and privacy concerns. It is crucial to develop robust security mechanisms that can protect AI systems throughout their lifecycle, from development and training to deployment and inference.

Adversarial Attacks on AI

Adversarial attacks are a major concern in the security of AI systems. These attacks involve the deliberate manipulation of data or the AI model to deceive or exploit the system. Adversaries can perturb input data, inject malicious code, or manipulate training data to achieve their malicious objectives.

One prominent example of an adversarial attack is the *adversarial example* attack. Adversarial examples are carefully crafted inputs that appear normal to humans but can mislead AI models. By making subtle modifications to the input data, attackers can cause the AI model to misclassify them, leading to potentially catastrophic consequences. Adversarial attacks pose significant threats in various domains, including autonomous vehicles, facial recognition systems, and natural language processing applications.

Game Theory in Adversarial Settings

Game theory offers valuable insights and techniques to analyze, understand, and mitigate adversarial attacks on AI systems. By modeling the interactions between the attacker and defender as a game, we can study their strategic decision-making and develop effective defensive strategies.

In the context of AI security, game theory can be applied to analyze and design robust AI models. One approach is to model the interaction between the attacker and the defender as a *two-player, zero-sum game*. The AI model developer acts as the defender, while the attacker seeks to exploit vulnerabilities in the system. The defender's goal is to minimize the attacker's impact, while the attacker aims to maximize the damage.

Strategies and Countermeasures

To defend against adversarial attacks, several strategies and countermeasures can be employed. These can be categorized into pre-emptive, reactive, and proactive measures.

1. *Pre-emptive Measures*: Pre-emptive measures focus on fortifying the AI system before any attack occurs. This includes robust model training techniques, such as adversarial training, where the model is trained with adversarial examples to enhance its resilience. Another approach is the use of anomaly detection algorithms to identify abnormal data patterns or behavior.

2. *Reactive Measures*: Reactive measures aim to detect and mitigate adversarial attacks as they occur. These include real-time monitoring and anomaly detection during inference. Techniques like input sanitization, where potential adversarial inputs are modified or discarded, can help prevent successful attacks.

3. *Proactive Measures*: Proactive measures involve staying ahead of adversaries by continuously updating and evolving AI models and defenses. This can include regular vulnerability assessments, model retraining, and incorporating new defenses based on emerging threat intelligence.

Ethical Implications

The deployment of AI systems in security-critical domains raises important ethical concerns. It is essential to consider the potential consequences and implications of the security measures employed. Some ethical considerations include:

1. *Bias and Discrimination*: When developing security measures, it is crucial to ensure that the implemented defenses do not introduce bias or discrimination against certain groups of people. Biases in data or model development can lead to unfair treatment or exclusion.

2. *Privacy and Surveillance*: AI security measures might involve collecting and analyzing large amounts of data, potentially compromising privacy. Balancing the need for security with individual privacy rights is of utmost importance.

3. *Transparency and Accountability*: Security measures should be transparent, allowing for scrutiny and accountability. AI developers should be able to explain the decisions made by their models and the security measures employed.

Real-World Applications

The application of security and game theory in AI extends to various real-world scenarios. Let's explore two notable examples:

1. *Autonomous Vehicles*: Autonomous vehicles heavily rely on AI technologies for navigation, object detection, and decision-making. Adversarial attacks targeting these systems can have life-threatening consequences. By applying game theory, security measures can be developed to protect autonomous vehicles from attacks that manipulate their sensors or compromise their decision-making algorithms.

2. *Cybersecurity*: AI is increasingly employed in cybersecurity to detect, prevent, and respond to cyber threats. By incorporating game-theoretic approaches, AI-based systems can anticipate and

counter adversarial attacks in real-time, resulting in more robust and effective cybersecurity defenses.

Further Reading and Resources

To deepen your understanding of security and game theory in AI, here are some recommended resources:

- "Adversarial Machine Learning" by Battista Biggio et al. (2018)
- "Game Theory for Security and Risk Management" by Stefan Rass et al. (2019)
- "Artificial Intelligence Safety and Security" by Roman V. Yampolskiy (2020)

Additionally, exploring academic journals such as the Journal of Artificial Intelligence Research (JAIR) and the Journal of Cybersecurity can provide valuable insights and the latest advancements in the field.

Conclusion

Security is a critical aspect of AI, and game theory offers powerful tools to analyze and address adversarial attacks. By understanding the principles and techniques of security in AI, we can devise robust defense mechanisms and navigate the ethical implications associated with them. As AI continues to evolve, the marriage of security and game theory will play an increasingly vital role in ensuring the reliability and trustworthiness of AI systems in our modern digital landscape.

Now that we have explored the fascinating realm of security and game theory in AI, let's move on to the final section of this book – the future of AI and game theory.

Ethical Implications of Game Theory in AI

As we dive into the world of Artificial Intelligence (AI) and its intersection with Game Theory, it becomes crucial to explore the ethical implications that arise from this amalgamation. AI-powered systems have the potential to make decisions on their own, and when coupled with the strategic nature of game theory, this raises important questions about responsibility, fairness, transparency, and the impact on society as a whole.

The Responsibility Dilemma

One of the key ethical dilemmas in AI and game theory is the issue of responsibility. When AI systems take part in strategic interactions, they make decisions based on the principles and objectives set by their designers. But who should be held accountable for the outcomes of these decisions? Is it the AI system itself, the programmers, or the organization behind it?

Assigning responsibility becomes particularly tricky in scenarios where AI algorithms learn and adapt over time. Reinforcement learning, a popular approach in AI, allows algorithms to learn from their interactions with the environment and optimize their decision-making processes accordingly. As a result, AI systems may make choices that their creators did not explicitly anticipate or endorse.

To address this dilemma, clear guidelines and regulations need to be established to ensure accountability and define the boundaries of AI decision-making. Algorithmic transparency and explainability are crucial in enabling a thorough understanding of AI decision processes, making it possible to trace back decisions and determine responsibility. Additionally, mechanisms for ongoing monitoring, auditing, and evaluation of AI systems must be in place to mitigate potential risks and ensure ethical behavior.

Fairness and Bias

Another ethical concern in AI and game theory revolves around fairness and bias. Game theory assumes rationality and fairness among players, but what happens when AI algorithms introduce biases into their decision-making processes?

Depending on the data used to train AI models, biases can mistakenly be incorporated into their decision-making processes, leading to unfair outcomes for marginalized groups or discrimination against certain individuals. Without careful consideration, game theory algorithms have the potential to perpetuate existing social biases or even amplify them.

Addressing bias in AI requires a multi-facetcd approach. Firstly, it is critical to ensure that training datasets are diverse, representative, and free from biases. Secondly, model evaluation metrics should explicitly measure and penalize unfair outcomes. Additionally, the design of AI algorithms should incorporate fairness constraints to guarantee equitable decision-making processes.

Furthermore, organizations that deploy AI systems need to enforce strict ethical guidelines and actively monitor their algorithms for any biases that might emerge over time. Transparent reporting and public disclosure of the decision-making processes can help guard against biased outcomes and enable external scrutiny.

The Human-Machine Relationship

As AI continues to advance, there is a growing concern about the impact of AI systems on the human-machine relationship. Game theory-based AI algorithms have the potential to outperform human players by exploiting their weaknesses and predicting their strategies. This raises questions about the fairness and integrity of human-machine interactions.

To address this concern, it is crucial to establish a proper balance of power and ensure that AI systems are designed to enhance human capabilities rather than replace them. Ethical considerations should drive AI development, with a focus on designing systems that augment human decision-making processes and empower users.

Additionally, it is essential to prioritize transparency in human-machine interactions. Users should have clear visibility into the decision-making processes of AI systems, enabling them to make informed choices and maintain control over the outcomes. Implementing human-in-the-loop frameworks, where humans have the final say in key decisions, can help mitigate potential ethical conflicts.

Unconventional Considerations: AI in Game Design

An unconventional yet significant ethical aspect to explore is the use of AI in game design itself. The application of AI algorithms can shape the user experience, influence player behavior, and even manipulate game outcomes. This raises questions about the ethical boundaries of using AI in these contexts.

Game designers must weigh the benefits of using AI algorithms to enhance gameplay against the potential negative effects. If AI is used to create addictive and manipulative game mechanics or to exploit players' weaknesses, ethical concerns become apparent.

To ensure ethical game design, transparency and informed consent are vital. Players should be aware of the AI elements in the game, their impact on the gameplay experience, and any potential data collection or privacy implications. Implementing ethical guidelines for AI-driven game design can help maintain a balance between creating engaging experiences and respecting players' autonomy.

Conclusion

As AI capabilities continue to advance and integrate with game theory, it is imperative to contemplate the ethical implications that arise. Addressing issues of responsibility, fairness, the human-machine relationship, and game design ethics are paramount to ensure AI-driven game theory aligns with societal values and benefits humanity as a whole. By adopting a proactive approach and integrating strong ethical frameworks into AI development, we can shape a future where AI and game theory work in harmony, maximizing benefits while minimizing potential harm.

The Future of AI and Game Theory

As we enter the era of artificial intelligence, the intersection of AI and game theory holds immense potential for shaping the future of both fields. Game theory has long been used to understand and analyze strategic interactions among rational agents, while AI has revolutionized decision-making and problem-solving capabilities. In this section, we will explore how AI can enhance game theory and pave the way for exciting developments in various domains.

Reinforcement Learning and Game Theory

One of the most promising approaches to integrating AI and game theory is through reinforcement learning. Reinforcement learning is a branch of machine learning that enables AI agents to learn optimal strategies by iteratively interacting with an environment and receiving feedback in the form of rewards. By combining these concepts with game theory, we can create intelligent agents capable of making strategic decisions in complex environments.

In game theory, the concept of equilibrium plays a crucial role in analyzing the behavior of rational agents. Traditional game theory assumes that each player has full knowledge of the game's rules and the strategies employed by other players. However, in real-world scenarios, this assumption may not hold, leading to suboptimal outcomes. Reinforcement learning can address this limitation by allowing agents to learn and adapt their strategies through experience, even in environments with incomplete or imperfect information.

Moreover, in multi-agent settings, traditional game theory often focuses on finding Nash equilibria, where no player can unilaterally improve their own payoff. However, in dynamic environments, these equilibria may not provide satisfactory solutions. Reinforcement learning algorithms can help agents search for better strategies that lead to better outcomes than traditional equilibria.

Algorithmic Game Theory

Algorithmic game theory is an interdisciplinary field that leverages algorithms and computational techniques to study strategic interactions. The combination of AI and algorithmic game theory opens up exciting possibilities for solving complex optimization problems, designing efficient mechanisms, and understanding economic and social dynamics.

The future of AI and game theory lies in developing algorithms and models that can handle large-scale, real-world scenarios. Traditional game theory assumes a small number of players and perfect knowledge of the game structure. However, AI techniques can allow us to tackle problems with a massive number of agents and uncertain dynamics. By applying scalable algorithms and distributed computing, we can analyze strategic interactions in social networks, financial markets, and other dynamic systems.

Furthermore, algorithmic game theory can enable the design of mechanisms to incentivize cooperation and promote fairness. By considering the computational aspects of game theory, we can design mechanisms that encourage desirable behaviors and discourage harmful actions. These mechanisms can be employed in various domains, such as resource allocation, online auctions, and climate change mitigation.

Ethical Implications of AI in Game Theory

As AI becomes more prevalent in game theory applications, it is crucial to consider the ethical implications that arise. AI systems can have unintended consequences, bias, and unfair impacts on different individuals and groups. Therefore, it is essential to assess the ethical dimensions of AI algorithms and ensure they align with societal values.

One key concern is the potential for AI agents to manipulate strategic interactions for nefarious purposes. AI algorithms could learn exploitative strategies that harm individuals or reinforce systemic inequalities. To address this, we must develop ethical guidelines and regulatory frameworks for AI and game theory applications to ensure that they adhere to fairness, transparency, and accountability principles.

Additionally, the question of AI's role in reducing human agency in decision-making processes arises. While AI can enhance decision-making capabilities, it is essential to strike a balance between automation and human involvement. By carefully designing AI systems that incorporate human preferences and values, we can create frameworks that empower individuals rather than replacing their judgment.

The Future of AI-Enhanced Game Theory

Looking ahead, AI-enhanced game theory has boundless opportunities for advancements. Here are some areas where AI and game theory can shape the future:

- **Policy Design and Analysis:** AI techniques can assist in designing policies that promote fairness, efficiency, and sustainability. By analyzing complex systems and simulating various scenarios, game theory augmented with AI can provide insights into policy design, cost-benefit analysis, and decision-making processes.

- **Smart Grids and Energy Management:** As the world seeks sustainable energy solutions, game theory can guide decision-making in energy management. AI algorithms can optimize energy distribution, balance demand and supply, and incentivize users to conserve energy through dynamic pricing mechanisms.

- **Healthcare and Clinical Decision Support:** AI-powered game theory models can aid in optimizing healthcare resource allocation, clinical decision-making, and treatment strategies. By considering patient preferences and resource constraints, these models can optimize outcomes and enhance the efficiency of healthcare systems.

- **Cybersecurity and Defense:** AI techniques can detect and respond to cyber threats, enhance network security, and mitigate risks in strategic domains. By analyzing adversarial behaviors and applying game theory, AI algorithms can adapt and counteract potential threats in real-time.

- **Social Dynamics and Opinion Formation:** AI-powered models can analyze social networks, track information diffusion, and predict opinion dynamics. By understanding the underlying mechanisms of social interactions, these models can provide insights into political campaigns, marketing strategies, and social influence dynamics.

As AI and game theory continue to evolve synergistically, the boundaries of what we can achieve are only limited by our imagination. Through responsible and ethical development, AI-enhanced game theory holds the promise of revolutionizing decision-making, shaping strategic interactions, and ultimately contributing to a more equitable and prosperous society.

Further Reading and Resources

To delve deeper into the future of AI and game theory, the following resources provide valuable insight and exploration:

- *Artificial Intelligence: A Modern Approach* by Stuart Russell and Peter Norvig.

- *Algorithmic Game Theory* edited by Noam Nisan, Tim Roughgarden, Eva Tardos, and Vijay V. Vazirani.

- *Reinforcement Learning: An Introduction* by Richard S. Sutton and Andrew G. Barto.

- *Superintelligence: Paths, Dangers, Strategies* by Nick Bostrom.
- *Weapons of Math Destruction* by Cathy O'Neil.

Remember, the future lies in our hands. By embracing the power of AI and game theory, we can shape a future where strategic interactions are more informed, fairer, and more conducive to cooperative outcomes. So let's play smart, not hard, and unlock the immense potential that lies at the crossroads of game theory and AI.

Index

-effectiveness, 292

a, 1–30, 33–35, 37–65, 67–76, 78–97, 99–103, 105, 106, 108–117, 120–139, 141–144, 146–172, 175–238, 241–245, 247–264, 266, 267, 269–271, 273–275, 277, 278, 280–283, 285–299, 301–308
ability, 29, 30, 40, 108, 115, 120, 121, 155, 175, 178, 179, 183, 185, 188, 192, 193, 222
academia, 1
acceptance, 235, 238
access, 34, 49, 136, 155, 162, 163, 168, 241, 243, 253, 254, 260, 298
accessibility, 133
account, 14, 15, 25, 63, 76, 84, 94, 96, 105, 110, 139, 144, 165, 180, 226, 229, 249, 252, 253, 259, 274, 294, 295
accountability, 21, 30, 300, 302, 306
accounting, 53, 60, 137, 245
accumulation, 175, 189
accuracy, 75, 251
acquisition, 189
act, 12, 14, 19, 25, 39, 41, 60, 75, 125, 132, 136, 152, 161, 162, 168, 195, 196, 210, 215, 218, 219, 223, 231, 233, 241, 244, 247, 249, 291, 292, 294
action, 1, 6, 25, 26, 39, 54, 68, 88, 122, 125, 127, 181, 191, 197, 201, 227, 278, 279, 295
actor, 33
ad, 226, 290
adaptability, 155, 185
adaptation, 44, 154, 189
adaption, 175
addition, 25, 46, 121, 130, 193
address, 45, 61, 64, 110, 116, 122, 130, 133, 177, 194, 197, 198, 200, 201, 203, 221, 223, 263, 267, 268, 281, 298, 301–303, 305,

306
adoption, 138, 140, 144, 155–160, 170, 171, 189, 191, 208, 272
advance, 291, 298, 303, 304
advancement, 22, 30
advantage, 2, 9, 20, 23, 25–28, 40, 58, 71, 83, 101, 110, 121, 131, 134, 177, 186, 187, 190, 211, 213, 216, 228, 229, 236, 238, 241, 254, 261–263, 273, 275, 287, 288
advertise, 95, 96
advertiser, 290
advertising, 95, 96, 215, 217, 225–228, 236, 290, 291
age, 6, 19, 137, 160
agency, 218–221, 306
agent, 6, 93, 219–221, 223, 277, 279–287, 296, 305
aggression, 5, 6
agreement, 8, 57, 109, 111, 215, 216, 218, 236
AI, 298, 301, 302
aid, 210
airline, 63, 274
airspace, 292
Albert-László Barabási, 147
Alex, 106
algorithm, 279, 287, 290
algorithms, 6, 29, 31, 71, 93, 130, 150, 163, 229, 275, 278–281, 286, 288–291, 293, 299, 300, 302, 303, 305,

306
Alice, 55, 83, 93, 116, 117, 131, 158, 186, 195, 273, 274
alignment, 221
alliance, 233
allocation, 44, 81, 106, 108–111, 114, 117, 119–123, 131, 132, 134, 177, 210, 226, 238, 250, 253, 257, 258, 266, 288–291, 305
alternative, 8, 16, 25, 46, 54, 121, 125, 150, 155, 177, 252, 260
altruism, 6, 184, 187, 190, 192–195, 198, 205, 208
Amos Tversky, 16, 43, 46
amount, 124, 235, 253
analysis, 12, 14, 17, 27, 28, 44, 45, 64, 68, 74, 75, 87, 94–96, 130, 135, 140, 141, 144, 145, 153, 166, 167, 180, 215, 230, 245, 247, 252, 260, 262, 263, 271, 272, 296
anchoring, 25, 245
Andrei, 252
anger, 15, 40, 53
animal, 17, 173, 177, 195, 198, 199, 202–204
Anna Machin, 211
announcement, 261
anomaly, 299
answer, 218
ant, 206, 296

Index

anthropology, 6
Antonio Damasio, 39
applicability, 13
application, 11, 42, 62, 69, 74, 75, 80, 93, 97, 123, 126, 128, 132, 148, 153, 181, 197, 204, 206, 218, 230, 233, 238, 245, 258, 259, 261, 266, 267, 269, 272, 273, 282, 287, 290, 300, 303
appreciation, 204
approach, 9, 10, 16, 17, 21, 42–44, 58, 82, 130–132, 153, 154, 178, 208, 232, 251, 252, 260, 263, 286, 296, 299, 302, 304
architecture, 61
area, 30, 39, 64, 164, 259, 287, 292, 293
army, 2
arrival, 105, 108
arsenal, 99
ascending, 258
ask, 249, 260
aspect, 47, 136, 139, 149, 164, 166, 197, 198, 219, 225, 228, 229, 233, 235, 301, 303
ass, 1–3, 7
assessment, 16, 18, 34, 44, 231, 264
asset, 230, 233, 248–251, 253, 259–261
asshole, 2
assumption, 14, 33, 46, 52, 53, 59, 96, 125, 134, 185, 241, 247, 250, 251, 305
asymmetry, 26, 46, 222, 250, 264, 266
attack, 294, 299
attacker, 294, 299
attempt, 20, 223
attention, 172
attorney, 55
attribute, 164
auction, 17, 63, 69, 118, 212–214, 235, 255–259, 290
auctioneer, 212
auctioning, 258
audience, 156, 168, 225, 226
auditing, 302
authority, 219, 231
automation, 295, 306
autonomy, 34, 304
availability, 25, 37, 48, 53, 153, 163, 245, 260
average, 105, 110, 120, 137, 164
aversion, 43, 47, 48, 53, 54, 210, 252
awareness, 31, 191, 225

back, 2, 8, 25, 40, 82, 95, 198, 302
backdrop, 202
background, 223
bacteria, 12, 187, 198
balance, 6, 9, 21, 40, 42, 64, 72, 129, 133, 154, 194, 214, 216, 231, 281, 287, 290, 294, 303, 304, 306
ball, 6

bank, 271, 273
bargaining, 8, 60, 64, 111, 113–115, 121–123, 132, 233, 234, 236, 237, 260
baseline, 47
basic, 3, 51, 129, 249
basis, 14, 190, 191
Bass, 159
battle, 184
behalf, 219
behavior, 1, 2, 5, 6, 8–12, 14–19, 22, 26, 27, 30, 34, 35, 43, 45–49, 53, 59–62, 64, 65, 74, 81, 93, 98, 99, 125, 127, 130, 134, 135, 137, 138, 140, 141, 144, 145, 151–154, 156, 157, 161–164, 168, 170–173, 178, 180–182, 186–194, 196, 198–212, 215–217, 219, 221, 223–232, 235–238, 242, 245, 247, 249–254, 259, 266, 267, 270, 271, 273–275, 282, 284, 292, 293, 296, 297, 299, 302–304
being, 1, 2, 9, 20, 21, 25, 27, 37, 41, 58, 77, 139, 155, 162, 177, 179, 184, 192, 197, 214, 231–233, 258, 266
belief, 162, 231
Ben, 106
benefit, 57, 73, 100, 113, 124, 131, 176, 186, 192, 200, 201, 203, 205, 216, 218, 235, 236, 293, 295
betrayal, 127, 163, 164
betweenness, 136, 168
beverage, 226
bias, 15, 25, 37, 47, 51, 53, 150, 245, 300, 302, 306
bid, 85, 118, 212, 213, 226, 235, 249, 255, 258, 260, 290
bidder, 118, 212, 213, 235, 255
bidding, 69, 128, 213, 214, 226, 235, 237, 255, 256
biologist, 6
biology, 1, 5, 11, 17, 74, 92, 93, 119, 157, 173, 176, 180, 181, 183–188, 191, 192, 194, 198, 202, 208, 288
bird, 200
bit, 1
blend, 1
blink, 259
blood, 193, 203
bluff, 29
bluffing, 26, 27
board, 3, 296
Bob, 55, 83, 93, 116, 117, 131, 158, 186, 195, 273, 274
bonus, 121, 210, 219
book, 3, 119, 141, 156
boss, 6
branch, 1, 12, 91, 255, 289, 304
brand, 21, 225, 226, 228, 230, 238
branding, 225–228, 238

Index 313

break, 1, 20, 57, 109
breakdown, 106, 127, 178, 199
breathing, 40
broker, 169
brood, 198
budget, 51, 225, 290
building, 20, 21, 102, 128, 139, 163, 171, 196, 231, 232, 264, 295
burden, 108, 109
business, 1, 2, 13, 21, 25, 26, 48, 55, 63, 64, 81, 97, 101, 105, 106, 111, 117, 119–121, 128, 131–134, 139, 210, 225, 227, 230, 233
buy, 22, 23, 69, 245, 250, 259
buyer, 260
buying, 8, 17, 212, 260

calculus, 259
call, 261
campaign, 166, 226, 227
capability, 155
capacity, 217
capital, 156, 162, 230, 250, 261, 266
car, 8, 17, 41
carbon, 57, 108, 133
care, 1, 198, 204
Carol, 116, 117
case, 24, 34, 44, 48, 70, 71, 89, 118, 181, 238, 258, 292, 293
caste, 206
caveat, 185
celebrity, 146
cell, 76, 80, 124

centrality, 135, 136, 141–145, 152, 154, 157, 168, 169, 262
century, 3, 4
certainty, 68
chain, 90, 128, 132, 134, 146, 169
challenge, 4, 7, 24, 59, 69, 111, 122, 126, 146, 153, 170, 185, 203, 213, 219, 222, 228, 250, 271, 290, 295
chamber, 150
chance, 70, 130, 186, 196
change, 3, 18, 57, 68, 69, 74, 108, 109, 111, 122, 124, 125, 131–133, 140, 167, 170, 171, 177, 179–181, 185, 191, 196, 253, 305
changer, 4
chapter, 157
characteristic, 115
charge, 55, 195, 216
charisma, 37
Charles Darwin, 175
cheater, 187, 200
cheating, 19–21, 27, 28, 178, 198–201, 203
check, 7, 10, 117, 172, 204
chess, 1, 29, 30, 70, 72, 84, 87, 90, 296
choice, 3, 14, 16, 17, 20, 25, 33–35, 41, 46, 48, 49, 51, 61–63, 75, 86, 92, 115, 176, 186, 200, 215, 229, 231
city, 45
claim, 223

Claire, 106
clarity, 40
class, 101, 289
classroom, 52, 168
cleaning, 178, 199, 295
client, 37, 199, 200
climate, 5, 18, 57, 108, 109, 111, 122, 125, 132, 133, 177, 191, 305
clock, 258
closeness, 136, 168
clustering, 152
coach, 30
coalition, 63, 101, 102, 106, 109, 110, 115, 117, 120–122, 166, 295, 297
code, 298
collaboration, 30, 31, 34, 53, 101, 103, 130, 132, 133, 135, 160, 162, 171, 197, 201, 295, 297
collapse, 270
collection, 304
collective, 6, 17, 20, 21, 53, 56–58, 63, 64, 101, 102, 109, 122, 125, 133, 137–140, 156, 161, 167, 181, 188, 190, 191, 197, 200, 201, 230, 263, 279, 281, 292, 293, 295, 296
college, 34, 223
collide, 214
collusion, 125, 178, 213, 214, 216–218, 274
colony, 206, 296

column, 80
combination, 143, 154, 167, 183, 253, 259, 281, 292, 305
commerce, 212
commitment, 127
committee, 224
communication, 9, 21, 53, 55, 101, 102, 127, 136, 211, 232, 233, 258, 295
community, 57, 264
company, 34, 84, 85, 95, 100, 121, 132, 133, 219, 224–226, 228, 243, 244, 249, 250, 261–263
competence, 223
competition, 9–11, 15, 17, 19–21, 62, 69, 81, 100, 121, 125, 130, 175, 177, 178, 180, 183, 185, 192, 214–218, 225, 226, 228, 230, 233, 258, 274, 292, 296
competitor, 9, 100, 228, 230
completion, 281
complexity, 16, 24, 34, 153, 157, 189, 191, 212, 250, 251, 288, 291
compliance, 219, 269
component, 231
composition, 101, 189
compromise, 8, 21, 102, 114, 300
computer, 6, 11, 71, 92, 93, 130, 131, 275, 288, 290

Index

computing, 279, 305
concentration, 229
concept, 1, 3, 5, 6, 8, 9, 11, 16, 33, 46, 47, 49, 51, 59, 67–69, 71–73, 75, 78, 82, 84, 88, 91, 95, 101, 102, 105, 108–111, 113, 115, 116, 118, 119, 121, 123, 129, 141, 154, 157, 162, 165, 168, 169, 175, 179, 183–185, 189, 193, 197, 202, 209, 212, 215, 218, 227, 228, 242, 249, 253, 260, 264, 275, 277, 286, 292, 295, 304
concern, 30, 298, 302, 303, 306
conclusion, 13, 34, 84, 96, 99, 201, 284, 288
conduct, 213, 231
confess, 55, 126, 161, 195
confessor, 161
confidence, 231
confirmation, 15, 53, 150
conflict, 10, 17–19, 69, 74, 132, 134, 161, 163, 170, 204, 233, 238, 271
congestion, 74, 292
connection, 34, 79, 146, 147, 243, 248, 251, 277
connectivity, 136, 141, 148, 152
connector, 136
consensus, 165, 171
consent, 304
conservation, 194, 206
consideration, 20, 30, 252, 302
construct, 13, 91, 92, 289

consumer, 48, 178, 217, 225–230, 235–238, 274
consumption, 61, 208, 281
contagion, 208
content, 150, 153, 156
context, 19, 20, 52, 59, 61, 79, 111, 139, 158, 171, 175, 176, 179, 183, 185, 188, 190, 192, 195, 196, 199, 201–203, 218, 219, 223–225, 227, 235, 242, 245, 253, 255, 260, 271, 278, 284, 290, 292
contract, 37, 84, 85, 210, 223, 235
contrast, 152, 287
contribution, 105, 108, 110, 120
control, 5, 79, 120, 132, 169, 274, 294, 303
convergence, 140, 289
conversion, 290
conviction, 126
cooperate, 6, 8, 20, 23, 57, 80, 92, 102, 120, 122, 125, 127, 132, 139, 161, 163, 171, 176, 180, 184, 186, 187, 190, 191, 194, 195, 200, 201, 203, 210, 216, 273, 292, 294, 298
cooperation, 2, 5, 6, 9–12, 20, 21, 53, 57, 58, 61, 63, 64, 69, 93, 101–103, 111, 113, 123, 125–134, 139, 141, 160–164, 169–172,

175–178, 181, 184, 186, 187, 189–201, 203, 204, 206–208, 215, 216, 218, 221, 230, 231, 233, 271, 273, 280, 284, 287, 292, 305
cooperator, 200
coordination, 217, 250, 281, 294, 295, 297
core, 19, 39, 46, 109–111, 116, 117, 119, 123, 277
correlation, 219, 252
cost, 2, 20, 47, 85, 192, 197, 201, 205, 216, 223, 292
country, 108
couple, 51, 54, 269
course, 1, 39, 227
cover, 110, 111, 223, 293
coverage, 86, 263, 264
creation, 161
credibility, 230, 231, 233
credit, 223
creditworthiness, 223
crime, 55, 161, 195
crisis, 271
criterion, 51
cryptocurrency, 4
crystal, 6
currency, 2
curse, 213
customer, 54, 132, 230
cybersecurity, 6, 13, 130, 297, 298, 300, 301
cycle, 57

Dan Ariely, 212
Dana, 106
dance, 218
Daniel Kahneman, 16, 18, 43, 46
data, 29, 30, 39, 44, 60, 153, 253, 297–300, 302, 304
dating, 9, 224
David Besanko, 224
David Dranove, 224
David Easley, 147
day, 3, 4, 262
deal, 1, 8, 55, 126, 132, 161, 195, 211
dealership, 41
deception, 2, 19, 162
decision, 1–4, 6, 9–13, 15–19, 21, 23, 25, 27–31, 33–35, 37–49, 51–63, 65, 67, 69, 71, 72, 74, 75, 78–84, 86, 88, 89, 91–93, 95–97, 99–101, 103, 108, 111, 117, 119–123, 125, 126, 128, 131–134, 139–141, 151, 153–155, 157, 158, 161, 164, 167, 169, 171, 172, 178, 182, 185, 194, 202, 204, 206, 208, 210–212, 214, 215, 219, 222, 224, 226, 227, 229, 230, 233, 236–238, 241, 242, 244–250, 254, 255, 260–262, 266, 271, 273–275, 277–282, 284, 286, 288, 289, 291–300, 302–304, 306, 307

decline, 180, 243, 271
decrease, 179, 261
defect, 56, 58, 83, 124–127, 139, 161, 171, 176, 180, 181, 186, 187, 194, 200, 203, 273, 274
defecting, 57, 61, 130, 186, 200
defection, 58, 64, 93, 123, 125, 127, 130, 133, 134, 163, 171, 176, 181, 186, 192, 194, 196, 203
defector, 200
defender, 294, 299
defense, 301
definition, 67
deforestation, 194
degradation, 197
degree, 135, 136, 168, 215, 222, 241
deliberation, 33
delivery, 292, 295
demand, 24, 54, 216, 229, 250, 294
demographic, 137
density, 165
departure, 251
depletion, 194
deployment, 298, 300
depth, 81, 156
design, 34, 63, 93, 99, 119, 133, 140, 145, 153, 156, 159, 160, 163, 166, 171, 194, 197, 201, 208, 221, 223, 224, 257–259, 263, 266, 267, 269, 275, 280, 281, 289–291, 294, 302–305
designing, 6, 23, 61, 71, 130, 132, 156, 163, 169, 222, 224, 245, 263, 284, 288–291, 295, 296, 303, 305, 306
desire, 9, 53, 210, 219
detection, 297, 299, 300
detective, 1
determination, 69, 289
deterrent, 200
deterring, 201, 217
detriment, 125
development, 31, 87, 133, 134, 153, 161, 177, 187, 189, 288, 298, 300, 303, 304, 307
deviation, 69, 210, 290
dialogue, 21
difference, 179
differentiation, 215, 228, 229
diffusion, 135–138, 141, 151, 152, 157–160, 170, 180, 262
dilemma, 20, 26, 55, 63, 123, 125, 161, 169, 171, 195, 216, 228, 236, 302
dinner, 1
dioxide, 133
diplomacy, 134, 211, 233
dirty, 2
disagreement, 10
disaster, 44, 45, 293–297
disclosure, 267, 303
discovery, 257
discrimination, 300, 302
disease, 153
disincentive, 127

dispute, 162
disruption, 270
dissolution, 121, 138
distancing, 153
distress, 271
distribution, 11, 74, 75, 81, 102, 111, 120, 121, 123, 132, 165, 184, 210, 234
district, 55
dive, 2, 4, 7, 9, 13, 19, 22, 37, 55, 69, 72, 79, 82, 94, 97, 100, 108, 157, 168, 198, 211, 214, 218, 255, 277, 288
divergence, 175
diversification, 44, 175, 253
diversity, 140, 155–157, 211
diving, 10, 35, 122, 248, 262, 291
division, 60, 63, 102, 115, 226, 234
domain, 63, 237, 297
dominance, 155, 228
dove, 184
dream, 41
Drew, 252
Drew Fudenberg, 224
drink, 228
drive, 34, 123, 171, 191, 194, 202, 218, 227, 230, 303
driver, 12
drone, 280, 281, 293
drug, 132, 133
Duncan J. Watts, 147
duopoly, 228
dynamic, 24, 78, 94, 115, 129, 157, 189, 196, 212, 214, 229, 259, 287, 290, 291, 305

earning, 216
Earth, 175
ease, 8
echo, 137, 150, 151, 166, 169
ecologist, 197
ecology, 17, 177, 204, 205, 208
economic, 4, 18, 22, 24, 29, 58, 97, 99, 100, 125, 156, 161–164, 177, 178, 180, 194, 197, 201, 210, 218, 305
economist, 6, 248
economy, 23, 24, 259, 266, 269
edge, 54, 86, 230, 260, 262
education, 191, 222
effect, 15, 37, 47, 54, 208, 271
effectiveness, 96, 131, 153, 194, 227, 254, 274, 292
efficiency, 132, 136, 216, 241–244, 257, 259, 266, 287, 290–292
effort, 47, 49, 53, 129, 223
element, 70, 75, 86, 101, 129, 139, 162
elimination, 73
Elon Musk, 146
emergence, 22, 30, 138, 155, 161, 164, 170, 171, 175, 177, 178, 180, 182, 187, 192, 194, 196, 198, 205, 208, 229, 284
emergency, 296
Emile Borel, 3
emission, 108
empathy, 53

Index

emphasis, 102
employee, 121, 210
employer, 210, 222
end, 94, 96, 228
endeavor, 102
endowment, 47
energy, 61, 184, 208, 281, 294
enforcement, 200, 232, 233
engagement, 150
enhancing, 155, 233, 273, 297
enjoyment, 34
entity, 230, 277, 282
entry, 124, 180, 215, 229
envelope, 212
environment, 6, 22, 58, 94, 141, 163, 175, 178, 179, 183, 185, 202, 255, 277, 279, 282, 287, 302, 304
epidemiology, 153, 159
equation, 77, 79, 259
equilibria, 23, 74, 75, 119, 134, 180, 181, 245, 248, 279, 281, 286, 289, 290, 292, 296, 305
equilibrium, 3, 5, 8, 14, 63, 67–69, 72–75, 77, 78, 80, 100, 118, 154–157, 170, 177, 183, 199, 203, 206, 218, 229, 242, 245, 249, 253, 271, 286, 290, 296, 304
equity, 102, 109, 211
era, 304
Eric S. Maskin, 224
Ernst Zermelo, 3
error, 297
essence, 83

establishment, 161
estimate, 248, 262, 278
estimation, 213
evacuation, 296
evaluation, 39, 53, 230, 302
event, 18
evidence, 15, 53, 55, 60, 126, 161, 195
evolution, 3–6, 12, 93, 125, 154–157, 170, 173, 175–181, 184, 185, 187–193, 195, 196, 201–208, 211
evolving, 3, 12, 31, 137, 153, 185, 230, 270, 300
exam, 34, 93
example, 5, 8, 9, 15, 20, 21, 24, 26, 29, 34, 37, 39, 41, 42, 45, 46, 51, 53, 55, 60, 61, 63, 64, 67–70, 77, 80, 81, 83, 87–89, 92, 95, 101, 105, 106, 115, 117, 118, 120, 125, 127, 130, 132, 133, 137, 152, 153, 159, 161–163, 168–170, 177–181, 184, 187–190, 193–197, 199–201, 203, 205–207, 211, 213, 219, 222, 223, 226–229, 243, 245, 251, 260, 261, 263, 271, 274, 275, 280, 289, 290, 292, 293, 295, 297
exception, 23
exchange, 197, 266
excitement, 40, 41

exclusion, 300
execution, 227
exercise, 109
exhaustion, 197
exhibit, 45–47, 64, 134, 152, 175, 203, 205, 206, 242, 243, 247
exist, 135, 137, 256
existence, 22, 75, 162
exit, 180
expectation, 193, 199
expense, 176, 219, 274
experience, 23, 41, 192, 224, 297, 303–305
experiment, 137
expertise, 34, 37, 101, 120, 121, 132, 255, 294
explainability, 302
explanation, 28, 189, 216
exploitation, 23, 163, 176, 201
exploration, 3, 22, 24, 85, 155, 206, 212, 251, 281, 294, 307
exposure, 150, 226, 254, 260, 261
expression, 190
extent, 136, 154, 156
extinction, 177
exuberance, 249
eye, 172, 259

face, 2, 29, 49, 54, 55, 57, 63, 69, 70, 74, 80, 138, 139, 161, 170, 177, 183, 191, 192, 194, 200, 228, 262, 266, 271
fact, 12, 155, 195
factor, 120, 162

failure, 41, 181, 211, 262, 271
fair, 2, 19, 20, 23, 24, 102, 105, 106, 108, 110, 111, 113, 114, 116, 120, 132, 198, 210, 214, 217, 219, 234, 247, 248, 258–261, 266, 270, 275, 291
fairness, 2, 19–21, 30, 60, 64, 71, 102, 109, 111, 133, 162, 207, 209–212, 234, 237, 238, 257, 267, 269, 280, 293, 297, 302–306
faith, 236
fallacy, 47
family, 162
farmer, 197
fashion, 3
fast, 255
favor, 27, 28, 121, 192, 205, 234
fear, 15, 40, 53, 162
feasibility, 109, 110
feature, 101
feed, 172, 193
feedback, 6, 127, 155, 189, 286, 304
feeding, 198, 205
feeling, 40
festival, 226
field, 3, 4, 6, 22, 44, 58, 62, 67, 81, 93, 148, 153, 154, 186, 208, 230, 244, 252, 259, 261, 273, 281, 290, 294, 297, 305
fight, 184
filter, 166

Index 321

finance, 44, 48, 62, 222, 223, 244–248, 250, 251, 255, 257–259, 261, 275
financing, 132
finding, 1–3, 5, 9, 72, 103, 111, 193, 203, 205, 207, 278, 280, 289, 305
firm, 74, 215, 216, 218, 235, 236, 274
fish, 178, 199, 200
fit, 22
fitness, 155, 175–183, 186–188, 190, 192, 193, 198, 199, 202, 205, 206
fixing, 21, 217
flexibility, 44, 131, 185
flow, 30, 74, 135, 136, 148, 152, 169, 249
focus, 47, 101, 103, 119, 257, 289, 299, 303
following, 37, 56–58, 69, 70, 85, 88, 93, 95, 99, 114, 122, 131, 180, 211, 224, 247, 267, 278, 307
food, 198, 296
foraging, 204
force, 26, 175, 195
forgiveness, 10, 196
form, 17, 23, 86, 88–90, 101, 105, 109, 111, 119, 120, 122, 137, 148, 151, 152, 166–169, 171, 177, 179, 187, 193, 199, 210, 230, 233, 242, 295, 304
format, 213
formation, 11, 18, 63, 101, 119–123, 137–141, 148, 150, 152, 163–167, 171, 180, 194, 197, 231, 233, 245, 249, 295, 297
foster, 21, 31, 53, 57, 127, 128, 133, 134, 162–164, 171, 191, 197, 207, 232
foundation, 3, 11, 33, 71, 111, 162, 191, 229
fragility, 162
framework, 6, 8, 11–14, 16, 19, 23, 25, 34, 43, 54, 58, 65, 79, 86, 90, 93, 111, 113, 128–130, 132, 133, 138, 151, 153, 158, 164, 167, 178–180, 192, 198, 199, 201, 202, 204, 206–208, 214, 225, 227, 229, 233, 235–237, 241, 244, 245, 248, 251–253, 259–262, 266, 271, 273, 275, 278, 281, 282, 286, 290, 292, 293, 295, 296
framing, 15, 37, 47, 48, 54
fraud, 23, 267
Frederick S. Hillier, 81
freedom, 61
frequency, 80, 177, 179, 181, 187, 200, 251, 258
friend, 7, 10, 35, 36, 38, 105, 106, 168
friendship, 135
frustration, 9
fuck, 1, 62

Fudenberg, 252
fuel, 41, 53
fulfillment, 39
fun, 27, 34
function, 43, 46, 79, 80, 115, 179, 278
functioning, 161, 163, 171, 197, 271, 273
fund, 253
fundamental, 11, 14, 46, 52, 67, 69, 72, 105, 109, 122, 135, 138, 139, 144, 149, 151, 162, 164, 176, 197, 229, 233, 242, 245, 249, 277, 282, 296, 297
fusion, 4, 293
future, 30, 31, 41, 44, 58, 95, 124, 127, 139, 153–155, 170, 175, 192, 198–201, 203, 231, 238, 245, 260, 261, 275, 288, 290, 294, 298, 304–308

gain, 2, 15, 20, 23, 25–28, 35, 39, 44, 47, 48, 54, 58, 67, 69, 70, 81, 89, 90, 97, 99, 101, 111, 119, 121–123, 126, 131, 139, 140, 145, 151–153, 155–157, 160, 163, 167, 171, 172, 180, 182, 191, 194, 196, 198, 201, 202, 204, 208, 213, 216, 218, 224, 226, 228–230, 235, 236, 238, 241, 242, 244, 247–249, 251, 252, 263, 271, 287, 288
game, 1, 2, 4–22, 24–31, 33–38, 42, 44–46, 52–55, 58–65, 67–73, 75, 76, 78–97, 99–103, 105, 108–112, 115, 116, 118, 122–126, 128, 129, 131–135, 138–141, 146, 151, 153, 154, 157–160, 163–173, 175–180, 182–192, 194–196, 198–209, 211, 212, 214, 215, 218, 219, 221, 225–233, 235, 237, 238, 241–255, 258–275, 277–308
gameplay, 26, 297, 303, 304
gaming, 2, 4, 22–24, 27, 86
Garrett Hardin, 197
gathering, 26, 27
generation, 176, 179, 192, 202
genius, 1
Gerald J. Lieberman, 81
getaway, 120
giving, 26, 110, 130, 194, 215
glance, 209
goal, 14, 101, 102, 161, 195, 202, 212, 216, 252, 258, 277, 287, 289
good, 20, 21, 49, 51, 64, 125, 129, 162, 207, 212, 230, 233, 236, 255, 278
gossip, 168
governance, 23, 121
government, 45, 85, 122, 210, 258

Index 323

grail, 4
greed, 25
grid, 294
grooming, 198
ground, 161
group, 8, 9, 18, 55–58, 101, 105, 120, 121, 139, 149, 151, 186, 189, 198, 199, 203, 219, 280, 293, 295
growth, 41, 125
guide, 19, 21, 39, 132, 133, 153, 169, 266, 267, 293
guy, 5

hand, 3, 16, 21, 25, 26, 30, 42, 47, 53, 95, 119, 151, 155, 176, 178, 188, 215, 219, 222, 223, 241, 242, 244, 248, 250, 277, 279, 298
Hanif D. Sherali, 81
happiness, 15, 53
harm, 21, 163, 226, 293, 304, 306
harmony, 304
hawk, 184
hazard, 223, 266
health, 138, 144, 148, 150, 155–157, 160, 171, 262
healthcare, 18, 294
heart, 33, 67, 161, 175, 195, 207, 292
hedging, 44, 261, 262
help, 1, 2, 7, 8, 21, 22, 25–27, 39–41, 44, 48, 52, 80, 81, 86, 92–94, 98, 100, 109, 111, 116, 121, 129, 132, 133, 144, 156, 157, 159, 163, 166, 168–170, 172, 177, 180, 185, 190–194, 196, 198, 200, 208, 211, 219, 224–226, 229, 243, 245, 250, 254, 270–272, 285, 292, 299, 303–305
helper, 205
helping, 8, 39, 97, 99, 192, 205, 212, 225, 248, 296
Herbert Simon, 17, 49
herd, 242, 245, 249
herding, 254
heuristic, 37, 49, 59
hiding, 205, 263
hiring, 222, 224
history, 3, 162, 217, 232
holder, 259
home, 34, 51
homogeneity, 155
honesty, 19, 162
hope, 48
host, 178
household, 61
hub, 136
human, 1, 6, 12, 16, 29–31, 34, 35, 43, 46, 49, 59, 60, 62, 74, 128, 145, 149, 157, 164, 172, 177, 178, 187–191, 197, 198, 201, 207–209, 211, 212, 245, 250, 291, 293, 303, 304, 306
humanity, 304

hypothesis, 39, 241

idea, 14, 16, 17, 70, 105, 109, 110, 113, 129, 135, 138, 157–159, 164, 170, 179, 242
identification, 14, 17
identifying, 23, 74, 132, 152, 163, 166, 171, 185, 245, 262, 273, 275
identity, 225
image, 200, 225
imagination, 307
imbalance, 20
imitation, 155, 159, 188
impact, 15, 16, 18, 21, 24, 30, 35, 37, 39, 44, 51, 53, 55, 59–62, 75, 101, 102, 121, 123, 134, 136, 137, 140, 145, 151, 152, 156, 162, 163, 169, 175, 185, 210, 211, 214, 218, 221, 226, 230, 233, 238, 244, 249, 251, 252, 259–263, 279, 286, 297, 303, 304
implementation, 93
importance, 8–10, 12, 23, 41, 46, 84, 105, 120, 130, 135, 140, 141, 144, 152, 157, 166, 168, 210, 224, 229, 234, 245, 259, 274, 300
incentive, 3, 5, 68, 69, 72–75, 80, 109, 132, 194, 197, 216, 219, 221, 253, 281, 289
income, 210, 223, 261

increase, 8–10, 25, 27, 44, 48, 120, 150, 171, 175, 179, 192, 213, 216, 225, 228, 231, 235, 236, 243
indicator, 222
individual, 1, 8, 20, 21, 34, 43, 53, 54, 61, 63, 101, 108, 110, 111, 119, 126, 138, 139, 141, 149, 151–155, 157, 158, 160, 161, 165, 168, 170, 178, 179, 188–194, 197, 202, 229, 230, 241, 252, 253, 263, 266, 270, 278, 279, 281, 287, 293, 295–297, 300
induction, 26, 82–84, 88, 89, 93–97
industry, 63, 217, 226, 227, 233, 235, 274
inequality, 210
inference, 298, 299
influence, 6, 12, 15, 18, 25, 34, 35, 37, 39–41, 47, 48, 52, 53, 58, 59, 61, 62, 64, 120–122, 127, 134–139, 141–145, 148–154, 156, 158–161, 164, 167–169, 171, 172, 176, 180, 188, 189, 191, 207, 208, 211, 212, 215, 219, 221, 225, 229, 230, 233, 236, 238, 242, 244, 249–251, 253, 254, 290, 303

Index

information, 8, 9, 11, 12, 14–17, 20, 21, 26–29, 33, 34, 39, 47, 49–51, 53, 59, 61, 70, 74, 86–90, 95, 96, 127, 135–138, 141, 144, 148, 150–153, 155, 157, 163, 168–170, 185, 221–224, 232, 241–245, 247, 249, 250, 253, 254, 260–264, 266, 267, 275, 288, 291, 293, 305
inheritance, 188
injustice, 235
innovation, 31, 155, 157, 159, 197, 259, 281
input, 298, 299
insight, 31, 75, 123, 131, 169, 188, 307
instance, 21, 29, 39, 48, 53, 54, 64, 127, 132, 152, 161, 189, 194, 205, 207, 219, 229, 243, 245, 264, 288, 292, 293, 296
institution, 271
insurance, 44, 262–266
insurer, 264
integration, 30, 31, 153, 279–281, 293
integrity, 19, 266, 269, 303
intelligence, 12, 40–42, 71, 93, 119, 247, 273, 277, 295–298, 300, 304
interaction, 23, 160, 183, 187, 189, 206, 218–222, 229, 233, 235, 236, 245, 248, 253, 260, 273, 277, 296
interconnectedness, 136, 141, 271
interdependence, 94, 215, 280
interdisciplinarity, 191
interest, 5, 8, 12, 21, 33, 34, 40, 56–58, 125, 134, 148, 161, 164, 195, 197, 205, 210, 216, 218, 219, 233, 244, 248, 249, 273
interplay, 24, 39, 41, 155, 157, 171, 180–182, 187–191, 205, 209, 215, 288, 291
intersection, 31, 151, 172, 204, 244, 248, 277, 291, 294, 298, 304
intervention, 153, 156, 291
interview, 223
introduction, 187, 207
intrusion, 297
intuition, 31, 42
invasion, 183, 185
inventory, 81
investing, 127, 228, 253
investment, 10, 39, 44, 48, 161, 244, 245, 248, 252–255, 261, 262, 274, 275
investor, 9, 241, 245, 251, 252, 261
involvement, 306
isolation, 164
issue, 18, 263, 280, 302
item, 118, 212, 213, 255, 258
iteration, 278

Jackson, 252
James H. Fowler, 147
James Mark Baldwin, 189
Jean, 252
Jean Tirole, 224, 252
job, 39, 119, 222, 223
John, 261
John J. Jarvis, 81
John Maynard Smith, 204
John Nash, 3, 5, 73, 113
John von Neumann, 248
Jon Kleinberg, 147
journaling, 40
journey, 7
joy, 39
judgment, 25, 34, 39–41, 49, 53, 230, 306

karaoke, 106
Ken Binmore, 18
key, 4, 33, 47–49, 53, 72, 79, 97, 101, 109, 120, 122, 142, 152, 154, 163, 166, 169–171, 177, 179, 183, 185, 192, 205, 214, 219, 224, 227, 245, 252, 257, 271, 278, 295, 302, 303, 306
kick, 2, 7, 35
kin, 192, 193, 195, 206
kind, 3
kindness, 161, 192
knife, 2
knowledge, 16, 27, 28, 33, 38, 71, 87, 161, 166, 185, 188, 189, 191, 197, 224, 229, 262, 282, 292, 295, 305

labor, 132
laboratory, 60
lack, 126, 127, 150, 195
landing, 292
landscape, 24, 31, 185, 212, 221, 225, 248, 258, 275, 301
lane, 287
launch, 95, 262
laureate, 17, 49, 73
lead, 9, 12, 14, 15, 17, 20, 21, 25, 31, 34, 35, 37, 39–42, 47, 49, 52, 53, 56, 59, 64, 70, 77, 82, 128, 132, 137, 150, 155, 171, 189, 190, 197, 199, 212, 224, 226, 229, 230, 233, 236, 242, 245, 260, 264, 277, 279, 288, 300, 305
leader, 236
learner, 279, 280
learning, 6, 29–31, 79, 93, 114, 154–157, 188–190, 247, 270, 277–281, 286–288, 290, 291, 297, 298, 302, 304, 305
left, 96
legislation, 121
lending, 223, 245, 246
lens, 19, 43, 126, 158, 195, 198, 207, 262, 290
letter, 137
level, 2, 10, 17, 44, 121, 159, 165
leverage, 121, 132, 134, 169, 294

Index 327

life, 1, 2, 7, 8, 10, 11, 17–19, 22, 27, 28, 38, 44, 45, 48, 52, 54, 55, 62, 63, 65, 75, 79, 82, 120, 122, 123, 133, 160, 175, 195, 204, 223, 224, 300
lifecycle, 298
light, 5, 12, 17, 56, 64, 71, 98, 125, 138, 166, 173, 177, 179, 181, 184, 194, 204, 205, 210, 227, 229, 249
likelihood, 48, 175, 231, 235, 250
limitation, 50, 134, 305
line, 216
link, 263
liquidity, 249, 262
list, 51
literacy, 150
livestock, 197
logic, 39, 46
loop, 127, 155, 303
losing, 47, 48, 213, 245
loss, 25, 39, 43, 47, 48, 54, 67, 69, 70, 80, 97
lot, 172
low, 100, 181, 205, 210, 211, 213, 216, 235, 236, 263, 290
loyalty, 225, 228, 230
luxury, 41
lynxes, 205

machine, 6, 30, 31, 93, 247, 290, 291, 303, 304
magnitude, 16, 46

maintenance, 162, 177, 192, 206
majority, 121, 146
maker, 43
making, 1–4, 6, 9–19, 21–25, 27–31, 33–35, 37–49, 51–63, 65, 67, 69, 71, 74, 75, 78, 81, 82, 84, 86, 89, 91–94, 96, 97, 99–101, 103, 108, 111, 113, 117, 119–123, 125, 126, 128, 129, 131–134, 139–141, 143, 151, 153–155, 157, 161, 164, 166, 167, 169, 171, 172, 178, 182, 185, 194, 202, 204, 208, 210–216, 219, 224, 226, 227, 229, 230, 233, 237, 238, 241, 242, 244–250, 253–255, 260–263, 266, 271, 273–275, 277, 279–282, 284, 286, 288, 291–300, 302–304, 306, 307
management, 27, 44, 45, 64, 79, 81, 90, 117, 127, 128, 132, 134, 177, 194, 198, 201, 238, 245, 253, 261–266, 281, 294–297
maneuvering, 24, 27
manipulation, 23, 223, 267, 275, 297, 298
manner, 14, 102, 162, 165, 231, 294
mapping, 93, 294

marker, 39
market, 4, 5, 9, 17, 18, 21–24,
    27, 39, 44, 61–63, 74,
    81, 92, 95, 121, 125,
    161, 177, 180, 185,
    214–216, 218,
    225–230, 236, 238,
    241–246, 248–251,
    253–255, 257–260,
    262, 266–271,
    273–275, 287, 291,
    297
marketing, 48, 62, 138, 144,
    148, 150, 153,
    155–157, 160, 164,
    166, 167, 170, 171,
    222, 227–230, 235–237
marketplace, 214, 227, 230
marriage, 301
mass, 138
mastermind, 38
mastery, 29
matching, 119
material, 175, 179, 219
mathematician, 73, 113, 248
mating, 189, 203, 204
matrix, 56, 67, 76, 80, 95, 96,
    124, 158, 180, 181,
    184, 186, 194, 205
matter, 3, 100, 209
Matthew O. ", 252
Matthew O. Jackson, 156
maximization, 257, 290
meal, 203
mean, 51, 252, 253
meaning, 59, 71, 155, 250
means, 20, 25, 39, 55, 72, 115,
    192, 210

measure, 120, 135, 136, 141,
    175, 179, 302
measurement, 143
mechanism, 130, 139, 177,
    190, 192, 193, 196,
    199, 200, 203, 207,
    222–224, 231, 255,
    263, 289–291, 295
media, 64, 140, 150, 153,
    155–157, 166, 172,
    231
meme, 157
memory, 39
mentality, 242
message, 226
method, 212
microstructure, 249, 251
mind, 4, 7, 35, 254, 262
mindedness, 150
mindfulness, 40
mindset, 2
minefield, 8, 36
minimax, 69–72, 75
minimum, 51, 124
misconception, 11, 13
misinformation, 137, 150
mispricing, 242, 245
mitigation, 44, 45, 305
mix, 181, 252, 282
model, 6, 14, 16, 17, 25, 33–35,
    46, 49, 64, 92, 113,
    138, 140, 158, 159,
    164–166, 176, 179,
    184, 186, 194, 195,
    202, 203, 222, 226,
    227, 229, 242, 253,
    254, 259–261, 271,
    279, 282, 287, 290,

Index

292, 293, 295, 297–300, 302
modeling, 4, 133, 152, 153, 166, 167, 180, 181, 192, 194, 202, 205–208, 229, 242, 245, 250, 252, 254, 273, 275, 288, 294, 296, 298, 299
modification, 188, 191
Mokhtar S. Bazaraa, 81
moment, 40, 202
money, 5, 8, 41, 47, 60, 63, 207
monitoring, 299, 302
monogamy, 189
monopoly, 214, 218
motherfucker, 6
motherfucking, 1
motion, 259
motivating, 218
motivation, 212
move, 6, 21, 26, 68, 70, 78, 82, 83, 95, 96, 125, 127, 129, 141, 186, 194, 232
movement, 160, 259
mucus, 199
multi, 6, 93, 277, 279–286, 295, 297, 302, 305
multiplayer, 24
multitude, 46
music, 3, 226
myriad, 35
Myron Scholes, 259

name, 11, 214
Nash, 3, 5, 68, 69, 72–75, 80, 100, 118, 119, 218, 242, 245, 249, 253, 279, 281, 286, 296, 305
naturalist, 175
nature, 11, 24, 57, 97, 101, 102, 128, 187, 193, 208, 213, 219, 236, 254, 271
navigation, 287, 300
need, 25, 58, 68, 73, 75, 76, 79, 100, 102, 106, 108, 111, 116, 117, 122, 133, 154, 192, 197, 210, 250, 252, 260, 261, 271, 275, 280, 287, 292, 293, 300, 302, 303
negotiating, 1, 17, 111, 121, 210, 234, 238
negotiation, 11, 17, 21, 28, 74, 84, 90, 101, 102, 111–115, 117, 131, 132, 134, 163, 238
nest, 296
net, 44, 192
network, 29, 64, 108, 120, 130, 135–141, 144, 148, 151–160, 162–165, 168–172, 262, 287
networking, 145
neuroscientist, 39
news, 27, 242, 253, 254
Nicholas A. Christakis, 147
night, 193
node, 86, 88, 91, 135, 136, 141, 152
non, 80, 110, 119, 194, 195, 219, 260
norm, 189

notion, 3, 16, 46, 109, 110, 210, 227, 245, 277
nourishment, 178
Nudging, 61
nudging, 61
number, 51, 62, 80, 96, 106, 126, 129, 135, 136, 138, 168, 193, 214, 229, 250, 274, 305
nutshell, 4

object, 300
objective, 79, 80, 252, 267, 295
obligation, 259
observer, 231
occupation, 137
off, 35, 109, 111, 113, 252
offer, 38, 39, 58, 60, 64, 86, 93, 108, 117, 123, 131, 157, 172, 188, 195, 209–211, 213, 215, 219, 234, 235, 238, 249
offering, 30, 31, 46, 131, 178, 223, 259, 263, 264
office, 8, 295
official, 45
offspring, 176, 179, 193, 205
oligopoly, 92, 130, 214, 215, 217, 218, 274
one, 20, 52, 55, 60, 67–70, 72, 80, 81, 85, 95, 97, 109, 113, 125–127, 129, 157, 161, 165, 183, 195, 200, 210, 214, 216, 219, 221, 222, 227, 228, 231, 236, 271, 274, 277, 289

opinion, 139–141, 164–167, 171
opponent, 26, 27, 68, 70, 73, 84, 125–127, 129, 130, 186, 194, 232, 287, 296, 297
opportunity, 2, 41, 48, 85, 94, 111, 122, 124, 186, 196, 231
optima, 155
optimality, 185
optimization, 74, 79, 296, 305
option, 17, 33, 43, 68, 70, 91, 126, 228, 259–262
order, 33, 75, 82, 85, 86, 94, 105, 108, 219, 249
organism, 175, 183, 192, 202
organization, 302
Oskar Morgenstern, 248
other, 2–4, 8, 9, 14, 16, 20, 21, 26, 28, 30, 39, 42, 47, 49, 53, 55, 60, 63, 64, 67–70, 72, 73, 77, 80, 81, 83–85, 87, 92, 94–97, 101, 106, 119–121, 125–127, 129, 130, 136, 137, 139, 141, 151, 152, 155, 161, 168–171, 176, 178, 179, 182, 183, 186–189, 194, 195, 199, 200, 203, 205, 211, 213, 215–217, 219, 221–223, 226, 227, 231–233, 236, 241, 242, 244, 245, 247–251, 253, 259, 260, 271, 274, 275,

Index 331

277, 279, 282, 285–287, 292, 293, 296–298, 305
outcome, 1, 11, 14, 18–20, 33, 37, 42–44, 51, 56, 70, 71, 73, 75, 82, 83, 85, 88, 89, 91–93, 95, 101, 113–116, 124, 131, 158, 165, 186, 195, 198, 245, 258, 271, 277, 286, 292, 295, 296
outcry, 212
output, 215, 216
overconfidence, 25
overexploitation, 128, 197
overgrazing, 197
overreaction, 242
overview, 157
owner, 48

pain, 47
pair, 68
panel, 217
panic, 271
paper, 25, 27, 67, 68, 70, 72, 76, 77
parcel, 10
parliament, 121
part, 2, 10, 63, 74, 160, 168, 279, 302
partnership, 9, 120, 121, 132
party, 3, 23, 26, 28, 34, 122, 132, 219, 221, 222, 231, 232, 263
past, 6, 29, 39, 127, 150, 156, 162, 163, 193, 200, 203, 230, 231, 238, 286, 290

pasture, 197
path, 168, 296
patience, 115
patient, 294
pause, 40, 42
Pavlov, 196
payoff, 20, 25, 56, 67–69, 72, 76, 80, 83, 95, 109, 110, 124, 158, 180, 181, 183, 184, 186, 192, 194, 201, 203, 205, 234, 249, 253, 260, 282, 295, 305
peace, 132
peak, 294
peer, 171
people, 2, 14, 25, 43, 46, 48, 53, 133, 137, 141, 150, 160, 169, 207, 210, 211, 244, 300
perception, 15, 210, 211, 226, 238
performance, 29, 129, 155, 219, 223, 280, 281, 293
persistence, 185, 205, 208
person, 8, 9, 33, 120, 137, 141, 162, 168
perspective, 7, 40, 46, 57, 192, 204, 205, 227, 238, 255
persuasion, 102, 165, 166
persuasiveness, 165, 166
Peter K. Smith, 211
phenomenon, 136, 137, 141, 146, 148, 149, 169, 170, 192, 210
philosophy, 33
piece, 244, 245
pizza, 116

place, 6, 102, 118, 138, 204, 212, 255, 302
plan, 18, 45, 68, 88, 92, 120
planet, 169
planner, 37
planning, 44, 45, 70, 120, 227
platform, 290
play, 2, 7, 12, 15, 17, 19, 21, 24, 25, 27–30, 39, 42, 46, 51, 53, 59, 60, 69, 72, 75, 79, 84, 90, 93, 94, 119, 122, 126, 127, 129, 134, 136, 143, 155, 156, 163–165, 171, 181, 191, 205, 208–210, 212, 213, 216, 218, 224, 225, 230, 236, 248, 255, 257, 266, 271, 287, 288, 290, 291, 301, 308
player, 2, 3, 5, 14, 15, 20, 21, 24, 26, 30, 31, 60, 67–73, 75–80, 82, 83, 86–89, 91, 94, 95, 97, 99, 100, 102, 105, 106, 108, 110, 113, 117, 118, 120, 121, 124, 127, 130, 186, 192, 194, 200, 203, 209, 214, 221–223, 227, 230, 231, 233, 242, 245, 249, 252, 253, 277, 285, 290, 296, 303, 305
playing, 2, 9, 11, 17, 20, 25, 28, 29, 31, 38, 68, 71, 76, 129, 141, 275, 287
plea, 55
pleasure, 47
plurality, 63
point, 16, 43, 47, 82, 83, 88, 91, 130, 154, 156, 227, 251
poisoning, 298
poker, 1, 26, 29, 70, 72, 296
polarization, 137, 140, 150, 166, 171
police, 195
policy, 18, 21, 44, 62, 63, 122, 138, 140, 166, 171, 194, 278
policymaking, 122
politician, 6
ponchos, 3
pool, 26, 101, 127, 263
popularity, 22, 155
population, 5, 175–177, 179–181, 183, 184, 186–191, 194, 195, 200–203, 205
portfolio, 252–255, 261
portion, 192, 193, 274
position, 121, 122, 141, 156, 158, 224, 230, 235, 261
positioning, 230
possibility, 124, 194, 231, 274
potential, 8–10, 15, 16, 28–31, 33, 34, 43–48, 54, 57, 58, 63, 70, 83, 91, 92, 94, 101, 102, 106, 109, 114, 122, 125, 130, 132–134, 136, 144, 158, 163–165, 169, 183, 195, 208, 212, 213, 215, 217, 222–224, 226, 228,

Index

231, 233–235, 238, 243, 244, 248, 251, 254, 258, 260–264, 272–274, 281, 285, 297–300, 302–304, 306, 308
power, 7, 8, 20, 24, 42, 64, 111, 119–123, 132, 134, 137, 151, 169, 206, 207, 215, 236, 260, 288, 303, 308
practicality, 41
practice, 115, 134, 146, 213, 247, 250, 251, 254
predator, 179, 204–206, 208
predictability, 247
prediction, 14
predisposition, 188–191
preference, 34, 43, 47
premise, 193
presence, 101, 134, 139, 187, 198, 201, 280, 292
presentation, 37
pressure, 52, 177
prevalence, 181, 203
prey, 179, 193, 204–206, 208
price, 8, 21, 51, 63, 74, 100, 180, 212, 213, 215–217, 227, 228, 235, 236, 242, 243, 248–251, 255–257, 259–261, 274, 275, 287
pricing, 4, 17, 48, 54, 62, 74, 81, 125, 130, 133, 185, 215, 216, 228, 229, 235, 245, 248–251, 259–262, 274, 290, 297

principal, 219–221, 223
principle, 19, 46, 52, 175, 183, 289
prison, 55, 80
prisoner, 26, 125, 161, 169, 171, 216, 228
privacy, 298, 300, 304
pro, 191, 208
probability, 33, 42, 43, 75, 80, 165, 179
problem, 16, 27, 79, 80, 99, 132, 191, 200, 219–221, 223, 250, 277, 289, 291, 293, 295, 304
process, 33–35, 37, 39, 40, 42, 45, 48, 49, 51, 52, 63, 73, 78, 83, 91, 95, 96, 111, 112, 121, 149, 154, 155, 157–159, 162, 165, 166, 175, 188, 212, 222, 224, 229, 233, 238, 242, 250, 252, 255, 274, 279, 281, 286, 290, 292, 293
product, 48, 51, 95, 113, 158, 160, 170, 215, 225, 228, 229, 235, 262
production, 81, 215, 216
productivity, 212
profit, 111, 117, 197, 235, 236, 242, 243, 260, 274
profitability, 133, 274
programming, 79–81
project, 8, 34, 37, 117, 219
prominence, 129, 141, 155
promise, 307
promotion, 10, 48

property, 51, 132
proportion, 176, 178–180
proposer, 63, 209–211
proposition, 47
prosecutor, 126, 161
prospect, 16, 25, 43, 45–49
provision, 125, 249
psychology, 1, 6, 11, 14, 17, 33, 48, 58, 71, 164
public, 44, 62, 125, 138, 140, 144, 148, 150, 155–157, 160, 164, 170, 171, 201, 207, 235, 238, 249, 260, 303
punishment, 170, 181, 190, 194, 195, 200, 201, 207, 208, 210
purchase, 41, 48
pursuit, 20, 139, 288
puzzle, 205

quality, 222, 226, 233, 236, 253, 254
quantity, 218
queen, 193, 206
question, 30, 105, 116, 126, 195, 207, 289, 306

race, 189
raise, 210, 216, 250
randomness, 68, 78
range, 4, 11–13, 44, 51, 99, 101, 109, 111, 121, 127, 130, 143, 155, 157, 178, 185, 187, 201, 205, 232, 280, 283, 291, 298
rate, 179, 291

rating, 162
ratio, 205
rationality, 11, 14–18, 34, 42, 46, 49–53, 55, 59, 62, 71, 74, 75, 91, 96, 134, 139, 185, 190, 209–212, 245, 250, 251, 254, 288, 302
reach, 74, 95, 114, 121, 133, 136, 137, 146, 156, 168, 169, 206
reading, 52, 126
reality, 11, 25, 29, 34, 50, 51, 62, 96, 127, 185, 247
realization, 103
realm, 54, 135, 156, 157, 163, 181, 189, 194, 198, 219, 233, 255, 263, 279, 298
reason, 46, 75
reasoning, 6, 16, 69, 75, 82, 83, 94, 96, 131, 133, 291
reciprocation, 199
reciprocity, 60, 64, 139, 162, 190, 196, 199, 203, 219
recognition, 129, 225
recombination, 188
recommendation, 223
reconciliation, 130
reduction, 108, 192, 263, 264
reef, 178
reference, 16, 43, 47, 48, 223
reflect, 40–42, 164, 241, 254
regard, 58
regret, 290
regulation, 23, 266–270
reinforcement, 29, 150, 277–281, 286, 287,

Index

290, 297, 298, 304
Reinhard Selten, 3
rejection, 211, 238
relatedness, 195
relation, 230
relationship, 9, 10, 24, 39, 127, 128, 153, 160, 191, 199, 219, 241, 266, 303, 304
relative, 16, 43, 46, 152, 176, 177, 193
relevance, 8, 57, 63, 110, 123, 135, 140
reliability, 231, 301
replicator, 178–182
report, 148, 243
reporting, 303
representation, 12, 16, 86, 120
representative, 302
reproduction, 175, 179, 188, 193, 197, 202, 206
reputation, 6, 21, 120, 127, 128, 130, 162, 170, 193, 196, 200, 219, 226, 230–233, 236–238
rescue, 293, 295
research, 14, 15, 53, 79, 81, 164, 171, 187, 204, 213, 247, 251, 259, 287, 290
researcher, 156, 195
resilience, 155, 162, 273, 299
resistance, 188
resolution, 17, 74, 132, 134, 162, 163, 204, 233, 238
resolve, 133, 134, 211
resource, 44, 64, 81, 108, 111, 117, 119–123, 127, 128, 131, 134, 177, 184, 194, 197, 198, 288, 291, 305
responder, 63, 209–211
response, 10, 100, 125, 127, 164, 189, 205, 226, 294, 296
responsibility, 30, 57, 133, 219, 302, 304
rest, 136
result, 69, 70, 150, 165, 187, 189, 242, 245, 260, 263, 271, 302
retaliation, 53, 130
retraining, 300
return, 193, 199, 248, 250, 252, 253
revelation, 289
revenue, 132, 257, 258, 290
review, 212, 259
revise, 85
reward, 181, 188, 196, 200, 231, 277–280
richness, 191
ride, 3
right, 9, 19, 168, 209, 226, 235, 259
rise, 2, 4, 22
risk, 6, 9, 16, 18, 25, 27, 42–45, 47, 53–55, 57, 124, 177, 219, 223, 231, 245, 248, 250, 252, 253, 259–266, 270, 271, 273
road, 287, 292, 293
Robert Axelrod, 204
Robert Axelrod's, 129
robot, 295, 297
robotic, 291

rock, 25, 27, 67, 68, 70, 72, 76, 77
Roger B. Myerson, 81
role, 2, 6, 10, 15, 17–19, 21, 24, 25, 30, 39, 41, 42, 47, 52, 53, 59, 60, 62, 64, 82–84, 102, 121–123, 127, 129, 132, 133, 135–139, 141, 143, 148, 149, 152, 156, 157, 161, 163–166, 170, 171, 175, 181, 191, 197, 202, 205–208, 210, 211, 215, 218, 225, 227, 228, 230, 232, 233, 238, 248, 251, 255, 257, 261, 263, 266, 273–275, 288, 291, 294, 297, 301, 304, 306
Rolf Schmittberger, 212
round, 126, 127, 129, 186
routing, 108
row, 80, 130
run, 20, 129, 181, 183

s, 1–11, 13, 15, 19–21, 24, 26–28, 30, 31, 34, 35, 37, 38, 41, 43, 46, 49–51, 54–59, 62–64, 67–75, 77, 79, 80, 82–84, 86, 89, 91, 92, 94–97, 99, 100, 105, 106, 111, 113, 115–118, 120, 121, 123–127, 129, 136, 137, 139–143, 154, 155, 157, 158, 160, 161, 168, 169, 171, 172, 175–177, 180, 181, 183, 186, 187, 189, 190, 192–195, 198, 200, 202–205, 207, 209–211, 213, 215–217, 219, 220, 222, 223, 226, 228, 230–234, 236, 243–246, 248–250, 252, 253, 255, 256, 258, 259, 261–264, 269, 271, 277–280, 282, 283, 286, 288, 290, 291, 294, 296, 297, 300, 305, 306, 308
safety, 44, 287, 292
salary, 17, 39, 63, 210
sale, 212
Sam Vecenie, 18
Sam Vecenie", 18
sanitization, 299
satisfaction, 14, 43, 132, 290
savvy, 7
say, 8, 9, 70, 146, 303
scalability, 288
scale, 153, 191, 208, 305
scandal, 244
scenario, 8, 28, 34, 37, 44, 45, 48, 57, 58, 71, 85, 93, 100, 108, 120, 122, 123, 128, 158, 161, 179, 186, 189, 216, 224, 227, 228, 233–235, 238, 243, 244, 251, 261, 262, 272, 273, 280, 285, 292

Index 337

scene, 3
schedule, 39
scheduling, 81
scholarship, 224
school, 8
science, 6, 11, 33, 63–65, 71, 79, 81, 92, 93, 130, 131, 164, 166, 171, 288, 290
score, 34
screening, 191, 221–224
scrutiny, 300, 303
search, 293, 295, 305
secret, 79
section, 3, 8, 11, 25, 35, 37, 39, 41, 42, 45, 52, 55, 62, 65, 67, 69, 72, 75, 78, 82, 93, 100, 101, 111, 115, 119, 123, 126, 128, 129, 131, 140, 141, 151, 153, 154, 157, 160, 164, 188, 198, 204, 209, 218, 225–227, 230, 233, 241, 244, 248, 252, 255, 259, 261, 262, 266, 270, 273, 277, 290, 298, 304
security, 133, 294, 297, 298, 300, 301
seeker, 222
seeking, 43, 47, 52, 236
seesaw, 67
segment, 226
segmentation, 230
selection, 5, 148, 173, 175–178, 188, 189, 192, 193, 195, 202, 205, 206, 223, 252–255, 263, 264, 266
self, 5, 12, 14, 17, 21, 29, 33, 34, 56–58, 109, 125, 134, 161, 164, 195, 197, 205, 210, 216, 218, 219, 291
selfishness, 187
selflessness, 192
seller, 8, 255, 257
sender, 223
sense, 2, 9, 41, 162, 219
sensitivity, 16, 44–46, 48
sentence, 55, 126, 161, 195
sentiment, 254
separation, 147, 169
sequence, 84, 86, 91, 129
series, 94, 232
service, 48, 212, 255
set, 1, 3, 51, 55, 68, 79, 80, 105, 109, 110, 135, 212, 216, 224, 228, 235, 236, 245, 250, 260, 286, 295, 302
setting, 3, 52, 55, 100, 117, 158, 228, 235, 279
shading, 213
Shane Parrish", 18
shape, 31, 35, 61–63, 120, 121, 138, 139, 141, 144, 166, 181, 188, 189, 198, 202, 204, 212, 219, 231, 269, 288, 291, 294, 303, 304, 306, 308
share, 26, 101, 110, 113, 137, 141, 152, 153, 156, 168, 169, 184, 192, 193, 198, 203, 215, 219, 225–228, 230,

232, 236, 287, 292
sharing, 34, 109, 116, 117, 119, 120, 132, 161, 176, 186, 193, 197, 204, 264, 266, 287
shit, 8
Shleifer, 252
shogi, 29
shot, 129
sign, 216
signal, 127, 221–224, 230
signaling, 127, 128, 217, 221–224
significance, 74, 83, 97, 122, 135, 141, 230
silent, 55, 126, 161, 195
similarity, 137
Simon, 49
simplicity, 131
situation, 8, 10, 11, 18, 25, 39–42, 48, 110, 121, 132, 133, 150, 186, 197, 205, 211, 222, 224, 234, 274
size, 101, 137, 153
skill, 223, 224
skillset, 94
sleeve, 36
smartphone, 18, 227
soccer, 285
socialization, 148
socializing, 34
society, 20, 21, 30, 57, 64, 133, 188, 191, 197, 258, 262, 291, 294, 307
sociology, 6, 17, 33, 164, 177, 180, 182, 184, 185, 194, 204, 207, 208
software, 291

solution, 17, 58, 79, 80, 102, 105, 108, 109, 111, 113, 116, 119, 132, 289
solving, 27, 29, 76, 77, 79, 80, 96, 132, 201, 291, 295, 304, 305
space, 24
speaker, 37
spectrum, 258
speculation, 254
speed, 136, 287
split, 234
spread, 44, 64, 135, 137–139, 144, 152, 153, 156, 157, 160, 164–167, 169–171, 177, 180, 187, 189, 191, 205, 208, 271
stability, 5, 11, 20, 72, 75, 109–111, 133, 177, 183–185, 193, 196–200, 203, 246, 266, 267, 269, 271, 273
stage, 3, 29, 94–96, 210, 215
stakeholder, 134
standoff, 3
standpoint, 41
Stanley Milgram's, 137
start, 72, 80, 89, 91, 95, 105, 120, 129, 168, 255
starting, 82, 91, 101, 106, 156, 251, 262
state, 3, 15, 68, 69, 72, 73, 75, 87, 154, 155, 242, 245, 253, 277, 278, 296
status, 41

Index   339

step, 25, 28, 29, 40, 82, 95,
    106, 165, 260
Steven Tadelis, 81
stock, 9, 39, 243, 245, 251,
    261, 274
storm, 22
strategy, 3, 5, 10, 13, 26, 29,
    44, 57, 68–73, 75–78,
    80, 83, 88, 95, 96, 100,
    101, 124–131,
    154–157, 170,
    177–181, 183–187,
    192–194, 196, 201,
    203, 205, 213, 218,
    222, 224, 225, 227,
    228, 232, 235, 242,
    243, 249, 253, 264,
    274, 281, 285–287,
    290, 296
strength, 166
stress, 272
strike, 211, 306
structure, 16, 17, 26, 79, 91,
    128, 135–141, 145,
    152–154, 156, 158,
    160, 161, 168, 177,
    187, 208, 214, 218,
    250, 274, 305
student, 34, 224
study, 5, 6, 17, 19, 24, 30, 34,
    48, 58, 60, 64, 69, 93,
    125, 135, 137–140,
    151, 153, 158, 160,
    163–165, 168–170,
    177, 178, 180, 185,
    187, 192–195, 197,
    201, 203, 204, 235,
    238, 244, 246, 258,
    259, 274, 279, 282,
    283, 288, 292, 299,
    305
subfield, 277
subgame, 3
subset, 109, 221
success, 9, 21, 25, 27, 34, 35,
    78, 82, 84, 95, 101,
    120, 121, 134, 156,
    175, 179, 181,
    185–189, 192, 193,
    195, 200, 202, 203,
    205, 211–214, 221,
    227, 230
sucker, 124, 181
suit, 159
sum, 2, 60, 63, 67–72, 75, 76,
    79, 80, 97, 99, 100,
    106, 117, 207
summary, 71, 251
supply, 5, 24, 90, 128, 132, 134,
    229, 294
surface, 31
surgeon, 294
surplus, 113
survey, 148
survival, 175, 179, 181, 183,
    185, 192, 193, 195,
    197, 198, 202, 203
suspect, 80, 126
sustainability, 198, 200
swarm, 292, 295, 297
system, 29, 178, 193, 206, 210,
    270–273, 282, 285,
    291–293, 298, 299,
    302

table, 105
tac, 70
tactic, 26

tailor, 150, 229
taking, 6, 14, 15, 25, 47, 63, 76, 82, 84, 89, 94, 96, 105, 138, 139, 164, 178, 180, 186, 226, 229, 253, 254, 259, 261, 274, 291, 294, 295
tango, 218
target, 108, 137, 156, 166, 225, 226, 236
task, 122, 142, 143, 219, 223, 252, 280, 281
taste, 3, 10
tat, 10, 21, 124, 126–131, 139, 178, 232
taxation, 23
teaching, 188
team, 101, 219, 280, 281, 293, 295
tech, 249, 250
technique, 6, 44, 79, 82, 96, 97
technology, 2, 22, 166, 261, 270, 291, 294, 298
temptation, 20, 21, 126, 127, 161, 181, 207
tendency, 46, 47, 152, 245
tennis, 84
tension, 20, 58, 63, 123, 125, 139, 161, 176, 190, 200
term, 9, 20, 21, 41, 57, 70, 126–130, 140, 179–181, 190, 193, 196, 199, 201, 207, 231, 233
territoriality, 184
territory, 202
test, 38, 60, 99
testing, 272

textbook, 126
theoretic, 54, 93, 139, 153, 163, 165–167, 193–195, 199–201, 249–251, 253–255, 259, 260, 262, 263, 266, 268, 269, 271, 272, 275, 281, 286–288, 292, 294, 296, 298, 300
theory, 1, 4–25, 29–31, 33–39, 42–49, 52–55, 58–65, 67, 69, 72, 75, 78–81, 84–86, 90, 91, 93, 94, 96–100, 102, 103, 105, 108, 109, 111, 119, 122, 123, 126, 128, 129, 131–135, 138–141, 151–154, 157–160, 164, 167–173, 175–179, 182, 185, 186, 188–192, 194, 195, 197–210, 212–215, 218–221, 225–231, 233, 235–238, 241–255, 258–267, 269–271, 273–275, 277–308
thing, 209
think, 8, 35, 39, 51, 109, 131, 209
thinking, 2, 6, 8, 18, 24, 27, 30, 39–42, 58, 59, 96, 102, 126, 129, 150, 235
Thomas C. Schelling, 18
thought, 41, 209, 218, 230, 253
threat, 273, 294, 300
threshold, 51, 138, 159
thumb, 59

Index

ticket, 63, 274
tie, 70
time, 3, 5, 10, 16, 31, 34, 42, 47, 49, 51, 52, 68, 70, 72, 83, 84, 93, 94, 109, 125–127, 131, 151, 155, 162, 165, 170, 172, 175, 179–181, 185, 189, 190, 193, 196, 198, 202–204, 208, 224, 231, 247, 260, 277, 289, 290, 293, 295, 297, 299, 301–303
timeframe, 259
timing, 26–28, 82–84
tissue, 178
Tit, 186, 187, 194
tit, 10, 21, 126–131, 139, 178, 232
today, 3, 191
toe, 70
tolerance, 43, 252
tool, 6, 11, 34, 35, 69, 71, 79, 81, 91, 94, 96, 99, 105, 108, 152, 176, 186, 203, 205, 215, 225, 230, 236, 242, 275, 294, 296
toolkit, 154, 204, 236, 255, 262
toolset, 24, 267
top, 214
topic, 122, 153, 290
total, 67, 80, 106, 110
track, 193
traction, 156
trade, 5, 23, 132, 161, 252, 275
tradeoff, 250
trader, 262, 273
trading, 2, 22, 242, 243, 249, 251, 259, 274, 275, 287
traffic, 30, 74, 281, 292, 293
tragedy, 125, 198
training, 31, 298, 299, 302
trait, 37, 189
transaction, 128, 250, 259
transfer, 115, 118, 119, 152, 263
transmission, 136, 153, 156, 157, 188–190
transparency, 30, 267, 269, 302–304, 306
transportation, 74, 292
trap, 213
treatment, 20, 300
treaty, 132
tree, 26, 86, 88–93, 260
trend, 157
trial, 297
trigger, 127, 128, 130, 131
trip, 36
trolley, 293
trust, 6, 9, 12, 20, 21, 40, 53, 63, 64, 102, 127–130, 133, 134, 139, 141, 160–164, 169–171, 201, 219, 230–233, 236–238, 271
trusting, 125
trustworthiness, 162, 231, 301
truth, 39
turn, 31, 82, 172, 179, 189, 223, 224
type, 67–69, 126, 200, 212, 222, 224

uncertainty, 9, 15, 16, 25, 29,

42–45, 53, 55, 59, 75, 78, 129, 171, 213, 232, 250, 266, 292
understanding, 1, 2, 5, 8, 11, 13, 15, 22, 24, 25, 27, 30, 31, 33–35, 37, 39, 45, 48–50, 54, 58–60, 62, 64, 65, 67, 69, 74, 78, 81, 82, 84, 90, 93, 97, 99, 100, 108, 111, 115, 119, 123, 128–135, 138, 140, 141, 144, 145, 148, 153, 154, 156, 157, 160, 162–164, 166–169, 171, 172, 175, 177, 178, 180–182, 184–186, 188, 191, 192, 198, 199, 201, 202, 204–206, 208, 211, 213–215, 221, 224, 225, 227–230, 232, 234, 236, 237, 241, 242, 244, 245, 247–249, 251, 254, 255, 259, 261, 263, 266, 269, 273–275, 278, 281, 286, 288, 291, 292, 294, 301, 302, 305
unfairness, 210
union, 132
universe, 24
university, 223
unpredictability, 26
updating, 165, 287, 300
uptake, 156, 171
use, 26, 28–30, 42, 44, 45, 49, 61, 79, 88, 92, 93, 127, 128, 160, 169, 221, 223, 224, 229, 230, 234, 236, 238, 243, 258, 261, 263, 274, 275, 286, 294, 295, 297, 299, 303
user, 150, 163, 231, 290, 303
utility, 14, 16, 17, 20, 25, 43, 45, 46, 52, 53, 72, 80, 91, 115, 117–119, 124, 165, 219, 242, 252, 253, 282, 290

vacation, 18
vaccination, 153, 156
vaccine, 156
valuation, 213, 235, 250, 251
value, 8, 9, 13, 16, 19, 27, 34, 43, 44, 46, 47, 76, 91, 102, 105, 106, 108–110, 115–117, 119–121, 164, 179, 213, 225, 245, 248, 258, 260, 261, 273, 278, 287
vampire, 193, 198, 203
variance, 252, 253
variation, 130, 175, 188
variety, 120, 144, 220
vehicle, 287, 293
venture, 101, 120
version, 124
viability, 199
vice, 67, 72
Vickrey, 213
victim, 51
victory, 27, 70
video, 3, 4, 93

view, 158, 192
vigilance, 205
virality, 156
visibility, 156, 303
volatility, 259
volunteerism, 194
von Neumann, 4
Von Neumann's, 3
voter, 165, 166
voting, 5, 17, 63
vulnerability, 272, 300

wage, 132
war, 100, 215, 228
warehouse, 295
warfare, 24
warning, 264, 273
water, 64, 127
way, 3, 4, 6, 8, 16, 24, 25, 38, 40, 45, 47, 61, 70, 110, 111, 120, 126, 158, 168, 179, 192, 198, 203, 218, 223, 231, 249, 288, 289, 291, 293, 304
wealth, 22, 23
weapon, 79
web, 204
weekend, 120
weight, 4, 46
welfare, 21, 119, 139, 217, 257, 259, 274
well, 20, 21, 38, 40, 58, 63, 127, 129, 136, 138, 139, 158, 166, 168, 178, 180, 183, 186, 193, 200, 213, 222, 250, 253, 263, 266, 280, 289, 295

Werner Güth, 211
whole, 21, 152, 179, 199, 206, 218, 294, 304
wildfire, 157
willingness, 15, 127, 150, 191, 223, 233
win, 118, 235
winner, 213, 255, 289
work, 8–10, 39, 88, 89, 91, 92, 100, 101, 105, 111, 126, 156, 163, 204, 212, 224, 255, 287, 304
worker, 206
workforce, 81
working, 101, 108, 132, 161, 195, 295
workload, 101, 295
workplace, 52
world, 2, 5, 12, 13, 21–25, 27, 30, 31, 33–35, 38, 41, 42, 45, 46, 49, 51, 55, 56, 59, 61, 62, 65, 69, 74, 75, 77, 78, 81, 84, 86, 89, 90, 92–94, 97, 99, 100, 103, 109, 110, 113, 117, 119, 121, 123, 125–131, 134, 136, 137, 140, 141, 144, 146, 148, 153–155, 157, 159, 163, 167–169, 172, 178, 181, 182, 185, 197, 198, 201, 202, 204, 208–210, 212, 214, 216, 218, 220, 221, 223, 225–227, 229, 230, 232, 238, 243, 246, 247, 250,

251, 255, 257–259,
261, 262, 264, 269,
270, 273, 280, 281,
284, 285, 288, 290,
293, 298, 300, 305

worth, 105, 213
wrasse, 200

Xavier Vives, 224

year, 55

Milton Keynes UK
Ingram Content Group UK Ltd.
UKHW021831041024
449101UK00012B/722